Pigeonpea Hybrids and Their Production

NIPA® GENX ELECTRONIC RESOURCES & SOLUTIONS P. LTD.
New Delhi-110 034

Pigeonpea Hybrids and Their Production

A. N. Tikle
Senior Scientist, Plant Breeding & Genetics
Rafi Ahmed Kidwai, College of Agriculture
Sehore 466 001 (M.P.)

K. B. Saxena
Ex-Principal Scientist & Program Leader (Pigeonpea)
International Crops Research Institute for the Semi-Arid Tropics
Patancheru 502 324 (Telangana)

H. S. Yadava
Director Research Services
Rajmata Vijayaraje Scindia Krishi Vishwa Vidyalaya
Gwalior 474 002 (M.P.)

NIPA® GENX ELECTRONIC RESOURCES & SOLUTIONS P. LTD.
New Delhi-110 034

NIPA® GENX ELECTRONIC RESOURCES & SOLUTIONS P. LTD.

101,103, Vikas Surya Plaza, CU Block
L.S.C. Market, Pitam Pura, New Delhi-110 034
Ph : +91 11 27341616, 27341717, 27341718
E-mail: newindiapublishingagency@gmail.com
web: www.nipabooks.com

For customer assistance, please contact
Phone: + 91-11-27 34 17 17
Fax: + 91-11- 27 34 16 16
E-Mail: feedbacks@nipabooks.com

ISBN: 978-81-19254-36-1

Composed and Designed by NIPA®.

प्रो. अनिल कुमार सिंह
कुलपति
Prof. Anil Kumar Singh
Vice - Chancellor

राजमाता विजयाराजे सिंधिया कृषि विश्वविद्यालय
रेस कोर्स रोड, ग्वालियर (म.प्र.) 474 002
Rajmata Vijayaraje Scindia Krishi Vishwa Vidyalaya,
Race Course Road, Gwalior (M.P.) – 474 002
Tel.: 0751 – 2467673, Fax : 0751-2464141, E-mail: vcrvsaugwa@mp.gov.in

D.O. No./VC/2014-15/2353
Date: 27/11/2014

Foreword

The world 'hybrid' cultivar excites farmers in the expectations of high yields and greater returns as the hybrid breeding technology has demonstrated quantum yield jumps in various cereals, vegetables and fruits. The efficacy of mass pollen transfer from male to female parent through air or insects to affect cross-fertilization plays an important role in commercializing the hybrids in different crops. In most of the food legumes, absence of cross-fertilization is the major bottleneck in exploiting hybrid vigour at commercial scale.

India, being the place of origin of pigeonpea, has a major share in its area and production. At present, the availability of proteins among the poor in developing world is less than one third of its normal requirements; and with growing population, the protein pulses considered as critical source of plant based proteins and essential amino acid for the people of the world in general and for developing countries in particular. It is a crop with huge potential that makes a significant contribution to the food and nutritional security of people.

The discovery of stable cytoplasmic male sterility system and breeding commercial hybrids in pigeonpea is a landmark achievement. This new hybrid breeding technology is capable of substantially increasing the productivity of pigeonpea and become a trigger for pulse revolution in the country. The development of CMS based hybrid – RVICPH 2671 has paved a new way for easy development and identification of high yielding hybrids.

The manual on 'Pigeonpea Hybrids and Their Production' will encourage researchers in adoption of this technology to achieve food security through productivity enhancement in Pigeonpea. The authors have made sincere efforts to enhance the awareness and understand the Pigeonpea hybrid breeding technology and its new facets.

Anil Kumar Singh

Dr. H.S. Yadava
Director Research Services

राजमाता विजयाराजे सिंधिया कृषि विश्वविद्यालय
रेस कोर्स रोड, ग्वालियर (म.प्र.) 474 002
Rajmata Vijayaraje Scindia Krishi Vishwa Vidyalaya,
Race Course Road, Gwalior (M.P.) – 474 002

Telefax : 0751-2467672
Email: drsrvskvv@rediffmail.com

No. DRS/2014/278
Date:24.12.2014

From the Desk of Director Research

Plant improvement has led to many high yielding, modern day crops, and it is continuing challenge to produce varieties with high yield, disease and pest resistant, drought and salinity tolerance in an environment friendly manner. It is noteworthy that many of the traits and genes that are necessary for crop improvement are still present in their wild relatives and the present day efforts have made possible to access these genes through various approaches.

Recent success in exploiting hybrid vigour through cytoplasmic male sterility system in pigeonpea offers optimism for a possible breakthrough in yield potential of food legumes. While popularizing high yielding CMS based hybrids in pigeonpea, an economically sound seed production technology is also a key factor to supply hybrid seed at affordable price. Further efforts need to be directed towards diversification and stability of cytoplasmic male sterility and genetic amelioration of its parental lines through insulation against key diseases and insect pests. Exploration and identification of heterotic cross combinations with climate specific or of widely adaptable restorers would possibly help in stability of pigeonpea hybrids. Efficient male sterility system coupled with fertility restoration is of prime importance in dissemination of such hybrids. Concerned efforts in this direction in the recent past resulted in development of stable CMS lines and hybrids.

For systematic pavement of research program by institutional efforts continuity in ingenuity makes possible in achieving the goal of success. The efforts made in this direction, by the authors of this manual, will surely guide the researchers in the improvement and development of new pigeonpea hybrids in future.

H.S. Yadav

Acknowledgements

Since last five decades pigeonpea scientists are constantly sharing the concern about plateauing productivity of the crop. The pioneer work done by ICRISAT and ICAR towards breaking the productivity barrier through hybrid technology is now appreciated by one and all; and the authors join in welcoming this breakthrough. The authors would like to thank ICRISAT for sharing this technology with Rajamata Vijayaraje Scindia Krishi Vishwa Vidyalaya; and this resulted in the release of the world's first commercial pigeonpea hybrid RVICPH 2671 by Madhya Pradesh Government in 2010-11.The authors would like to appreciate and acknowledge the dedicated research efforts of ICRISAT Pigeonpea Breeding team in developing an easy and affordable hybrid breeding technology.

This manual deals with various aspects of research and commercial production of pigeonpea hybrids; and most information and photographs were made available by ICRISAT, and the authors greatly acknowledge this support.

Authors are highly grateful to the Hon'ble Vice Chancellor Prof. Anil Kumar Singh *Rajmata Vijayaraje Scindia Krishi Vishwa Vidyalaya*, Gwalior, for his constant inspiration in the preparation and publication of this manual.

Authors are also thankful to Dr. Sandeep Sharma, Entomologist of R.A.K. College of Agriculture, Sehore for his valuable technical guidance and editing of the manuscript.

Authors

Contents

List of Tables

List of Figures

1

Introduction

Pigeonpea (*Cajanus cajan* L. Millsp.) is crowned as poor man's crop which provides much needed protein to farming families with least inputs. It is a versatile plant species which can grow successfully in a range of soil types, temperatures and photoperiods. Its deep root system enables it to overcome intermittent drought and other stresses. In India, the crop is grown annually on about 4.04 m ha (FAO, 2012) and the major pigeonpea growing states are Maharashtra, Andhra Pradesh, Karnataka, Madhya Pradesh, Uttar Pradesh, Bihar and Gujarat (Table 1). The high protein pigeonpea *dal* (decorticated dry splits) is widely consumed across the country in various cuisines. The annual national production of pigeonpea is about 2.56 m tones. This produce, however, is not sufficient to feed the ever growing population of the country, and hence, necessitates huge (500,000 tons) imports of this pulse. In spite of its importance in food security and sincere research efforts, its national productivity could never cross the bar of 800 kg/ha (Fig 1) since independence. Since last 50 years, ICAR has been very seriously pursuing the issue of genetic enhancement with huge investments and some out-standing pure line varieties with about 10% more grains over the best local varieties were developed. These achievements, however, were not enough to feed the nation with nutritious food and a major breakthrough in pigeonpea productivity was needed. Intensive national efforts in this direction could not succeed in raising the stagnant productivity level. Recently, an opportunity had come in our way when cytoplasmic nuclear male sterility (CMS) based hybrid pigeonpea technology was developed jointly by International Crops Research Institute for the Semi Arid Tropics (ICRISAT) and Indian Council of Agricultural Research (ICAR). The release of the world's first commercial hybrid RVICPH 2671 by Rajmata Vijayaraje Scindia, Krishi Vishwa Vidyalaya, Gwalior, Madhya Pradesh has already become a milestone in pulse breeding

(Saxena *et al.*, 2013). This hybrid has demonstrated huge (>40%) yield advantage in farmers' fields. This research development has given the breeders an option to break the decades-old yield plateau in pigeonpea. For realizing high yields from hybrids and reap the benefits of this technology, the availability of both quality seed and optimum agronomy are essential. Since in hybrids every year new crossed seeds are required for sowing, the production of large quantities of quality seed is the key for their success. In this book efforts are made to describe various methods of breeding new hybrid combinations and large scale seed production of promising hybrids and their parents.

Table 1: Area, production and yield of top 10 pigeonpea growing states of India

Rank	State	Area ('000 ha)	Production ('000 t)	Productivity (kg/ha)
1.	Maharashtra	1081.0	909.0	841
2.	Karnataka	666.0	369.8	555
3.	Madhya Pradesh	530.5	402.3	758
4.	Andhra Pradesh	481.0	250.0	520
5.	Uttar Pradesh	311.0	373.0	1199
6.	Gujarat	228.0	270.0	1184
7.	Jharkhand	187.8	199.6	1063
8.	Orrisa	150.2	114.0	
9.	Chhattisgarh	52.1	32.3	620
10.	Tamil Nadu	51.9	39.5	761

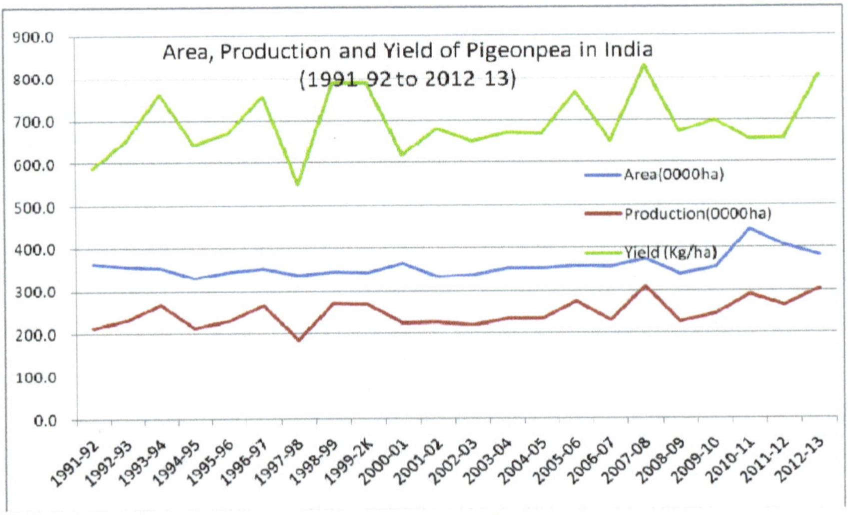

Fig. 1: Area, production and productivity of Pigeonpea in India

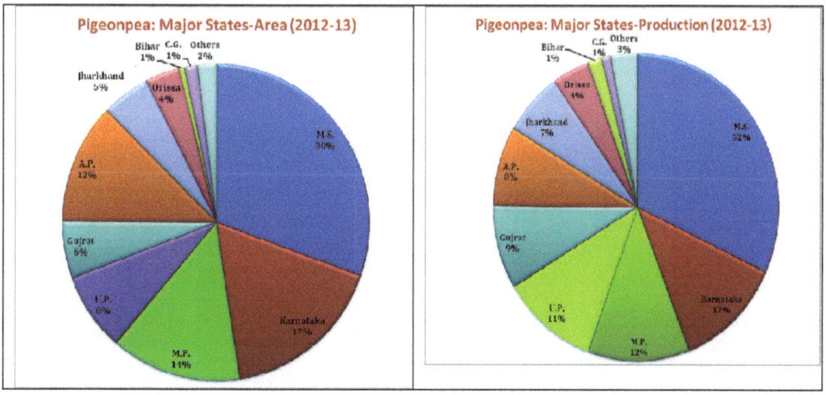

Fig. 2: Percent contribution of states in pigeonpea area and production in India

2

Male Sterility Systems

Male sterility is a situation where the male reproductive parts of a plant are either absent, aborted, or non-functional; and therefore, they fail to participate in the process of natural sexual reproduction. This situation generally arises due to some developmental defect that occurs at any stage of microsporogenesis (process of development or release of pollen grains) in the plants. Since male sterility is the manifestation of abnormal growth and development and can occur at any point of time, the action of genes controlling male sterility is also be variable and inconsistent across the genotypes. Based on the type of malfunctioning of androecium, the male sterility systems have been classified as structural (absence or deformity of anthers), sporogenous (defective microsporogenesis), and functional (failure of mature pollen to germinate). In addition, on the basis of genetic control mechanism of the male sterility, it has also been classified as genetic, cytoplasmic, and cytoplasmic nuclear (or genetic) male sterility. With respect to utilization of male sterility in crop breeding, it is essential that individuals with altered male fertility keep their female fertility intact. The fertilization of such plants takes place with pollen grains from other plant that is transferred with the help of external agencies such as wind, insects, or human beings.

Historically, the male sterile mutants had appeared naturally in the populations of cultivars and germplasm; but at that time their economic value was not recognized; and hence these were lost over a period of few generations. However, with the evolution of the concept of heterosis by Shull (1908) and subsequently by others, the potential benefits of male sterility in enhancing productivity of crops started gaining momentum. A breakthrough in this direction was achieved when Stephens and Holland (1954), for the first time, used the male sterile genotypes in hybrid seed production of sorghum. At about the same time, Jones and Emsweller (1937) also demonstrated the use of male sterility in hybrid seed production of onion. The male sterile trait may arise spontaneously through mutations or can be bred through induced mutagenesis, or by hybridization and selection. Among various types of male sterility systems reported in literature, cytoplasmic nuclear/genetic male sterility (CMS or CGMS) is the most important because of its potential use in commercial hybrid breeding programs; and in the last few decades, it has helped in encountering hunger specially in developing and under developed countries of tropics and sub-tropics.

2.1. Cytoplasmic-nuclear male sterility system

In cytoplasmic nuclear male sterility, the manifestation of male sterility is a consequence of interaction between the genomes of cytoplasm and nucleus. In this type of male sterility, dominant fertility restoring nuclear genes restore the male fertility of the hybrid plants. Further, depending on the type of fertility restoring gene and its strength, the expression of male fertility/sterility could be total or partial. Sometimes, such expressions are also affected by prevailing environmental conditions such as photoperiod, temperature, or both. The cytoplasmic nuclear male sterility has been used most extensively in hybrid breeding programs in a number of field crops. The complete hybrid system involves three distinct genotypes (Fig 3). These include 'A' line-the male sterile female line with sterile 'S' cytoplasm and recessive fertility nuclear alleles (*frfr*), the 'B' line is a maintainer of the female line; and it has fertile 'N' cytoplasm and recessive fertility nuclear alleles (*frfr*). This line when crossed with 'A' line, the entire progeny is male sterile. Third parent is designated as 'R' line and it contains dominant fertility restoring nuclear gene (*FrFr*). This line has ability to restore the male fertility of the hybrid plants produced by crossing it with 'A' line.

The natural occurrence of this form of male sterility system is low, because, it requires simultaneously mutations at two sites; one each in mitochondrial and nuclear DNA. According to Kaul (1988) so far only 46 plant species are credited

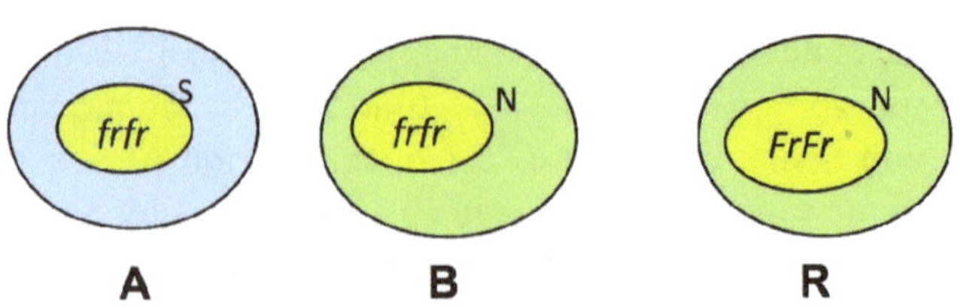

Fig. 3: Genetic structure of A, B & R line

to have produced this form of male sterility under natural conditions. Considering its value in commercial plant breeding in enhancing productivity of the crops, many attempts have been made to breed male sterile genotypes and the most common approach to develop CMS lines has been to integrate the genome of cultivated species with the cytoplasm of its wild relatives. This technology has been successfully used in cereals, oil seeds, legumes, and various other groups of crops. It is based on the concept of brining cytoplasmic and nuclear genomes of diverse origin together within a single genotype. This is achieved by crossing the wild relative of a crop as female parent with a cultivated line as male parent. This allows the transfer of entire cytoplasm of the wild species in the F_1 plants. These plants will have 50% nuclear genome of cultivated type. The entire genome of the cultivated line can be transferred in six back crosses using the cultivated type as recurrent parent. This can be achieved through inter-specific or inter-generic mating. In the inter-generic hybridization, the wild relatives from different genera are crossed as female parent with cultivated types, but due to large diversity, most of the times the crosses are not successful and it may require the use of embryo rescuing technology to raise the hybrid plant. The resultant hybrid plants may also show serious abnormalities due to altered meiosis. Such male sterile plants fail to survive in the absence of any maintainer. The inter-specific hybridization is the most successful approach in breeding cytoplasmic nuclear male sterility systems in a number of crops (Kaul, 1988). In this group of materials, the success from hybridization is relatively high due to closer genetic relationship between the two species and from such crosses, male sterile genotypes can be selected for back crossing.

In pigeonpea, so far, eight sources of cytoplasm are available which can induce cytoplasmic nuclear male sterility (Table 2). Among these, only two (A_2 and A_4) sources have been used in hybrid breeding programs. The A_2 CMS system has cytoplasm of *C. scarabaeoides* (Tikka SBS *et al.,* 1997) (Fig 4) and it is highly stable male sterility system with a number of fertility restorers available. However, the hybrid (GTH 1) derived from this CMS did not show high stability of fertility restoration under diverse environments, hence the hybrid was not accepted by seed growers. On the contrary, the A_4 CMS derived from *C. cajanifolius* (Fig 5) is highly stable (Saxena *et al.*, 2005) with respect to the expression of male sterility and fertility restoration. The hybrids (RVICPH 2671 and ICPH 2740) derived from this CMS system are stable and high yielding and hence released for cultivation in India. There are some other species useful for other important traits (Fig 6).

Table 2: List of eight CMS systems developed in pigeonpea

*CMS no.	Donor species	Recipient species
A_1	*C. serecious*	*C. cajan*
A_2	*C. scarabaeoides*	*C. cajan*
A_3	*C. volubilis*	*C. cajan*
.A_4	*C. cajanifolius*	*C. cajan*
A_5	*C. cajan*	*C.acutifolius*
A_6	*C. lineatus*	*C. cajan*
A_7	*C. platycarpus*	*C. cajan*
A_8	*C. reticulates*	*C. cajan*

*For references see Saxena *et al*. (2010)

Fig. 4: *Cajanus scarabaeoides*

Fig.5: *Cajanus cajanifolius*

Wild species	CMS system	Remarks	Reference
 *C. sericeus*	A1	CMS sensitive to temperature	Ariyanayagam et al.,1995
 *C. scara baeoides*	A2	Fertility restoration unstable	Saxena and Kumar, 2003
 *C. volubilis*	A3	Large variation in expression	Wanjari et al.,1999
 *C. cajanifolius*	A4	Stable, using in hybrid program	Saxena et al.,2005
 *C. cajan*	A5	Uses cultivated pigeonpea cytoplasm	Mallikarjuna and Saxena,2005
 *C. lineatus*	A6	-	Saxena (unpublished)
 *C. platycarpus*	A7	A new CMS using tertiary gene pool	Mallikarjuna et al., 2006
 *C. reticulatus*	A8	Searching fertility restoration	Saxena, 2013

Fig 6: Important *Cajanus* species for hybrid breeding program

Perfect restoration of male fertility of the cytoplasmic nuclear male sterility based hybrids is important for any hybrid breeding program. Once an 'R' line is crossed with 'A' line, the dominant *Fr* nuclear gene of 'R' line repairs the defective mitochondrial genome. It is believed that the *Fr* gene synthesizes certain proteins which are actively involved in repairing the damage in the mitochondrial genome; and this makes the hybrid plant male fertile. In most crops one or two dominant fertility restorer genes have been reported to control the pollen fertility (Kaul, 1988). In pigeonpea also, 1-2 dominant genes have been reported to control the fertility restoration of A_4 cytoplasm (Dalvi et al., 2008). Saxena et al. (2011) also reported that the pigeonpea hybrids with a single dominant gene were also fertile but they produced less quantity of pollens and were instable with respect to fertility restoration across diverse environments. On the contrary, the hybrids with two *Fr* genes were highly stable.

2.2 Other male sterility systems

Besides CMS, the male sterility system described above, three more male systems have been reported (Saxena *et al.*, 2010) and these are briefly described below:

Genetic male sterility: This is the most common form of male sterility found in a number of plant species in both, monocots as well as dicots. In this system, the male sterility is controlled by nuclear genetic factors and it is independent of cytoplasmic influences (Fig 7). In most cases, its expression is controlled by one or two pairs of recessive alleles, which segregate independently. However, a few exceptions are also found where the male sterility is reported to be controlled by one or two dominant genes (Kaul, 1988). The mutant male sterile plants may arise spontaneously carrying homozygous alleles (*msms*), and these will be lost if not maintained as heterozygotes (*Msms*). For this to happen, the male sterile mutants need to be pollinated with fertile homozygous (*MsMs*) or heterozygote (*Msms*) counterparts. In cases where male sterility is controlled by dominant alleles, its maintenance through reproductive means is very difficult. Saxena *et al.* (2010) compiled this information in pigeonpea (Table 3) and reported the presence of 11 sources of genetic male sterility. The allelic relationships among these have not been reported.

Table 3: List of genetic male sterile phenotypes reported

S. No.	Year	Description / remarks	Gene	Author*
1.	1959	Male sterility associated with female sterility	-	Deshmukh
2.	1977	Seven types of male sterile variants in germplasm	-	Reddy et al.
3.	1978	Translucent anthers	ms_1	Reddy et al.
4.	1981	Partial male sterile with sparse pollen	-	Saxena et al.
5.	1982	Photo-insensitive mutant	-	Dundas et al.
6.	1983	Brown arrow-head anthers	Ms_2	Saxena et al.
7.	1983	Recessive gene pair	-	Gupta, Faris
8.	1994	Linked to obcordate leaf	-	Pandey et al.
9.	1997	Single recessive genetic control	-	Verulkar, Singh
10.	2000	Single recessive genetic control	-	Wanjari et al.
11.	2001	Non-allelic to ms_1 and ms_2	Ms_3	Saxena, Kumar

For references see Saxena et al. (2010)

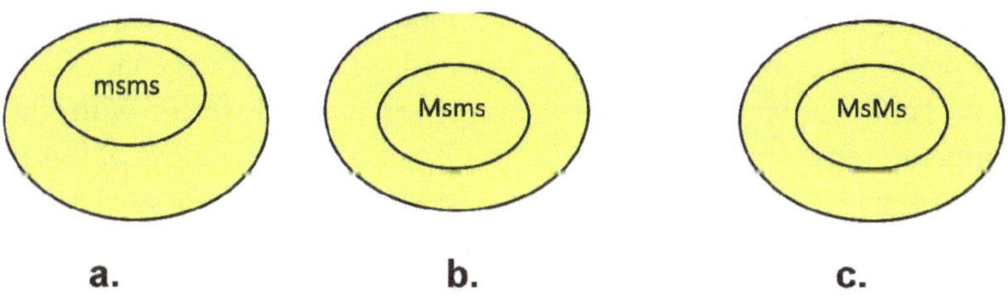

Fig. 7: a. Homozygous recessive b. Heterozygyous c. Homozygous dominant

Cytoplasmic male sterility: This type of male sterility is governed by cytoplasmic factors that contains defective mitochondrial DNA (Fig 8). This happens due to deleterious interactions of mitochondrial genes with those present in the nucleus. This type of cytoplasm is designated as 'sterile' (S); and it can originate spontaneously or through wide hybridization. Such plants do not produce fertile pollen grains because its nucleus also contains a pair of recessive non-restoring (*msms*) alleles. The cytoplasmic male sterility is maintained by the genotypes which carry 'fertile' (F) or 'normal' (N) cytoplasm and non-restoring recessive nuclear alleles.

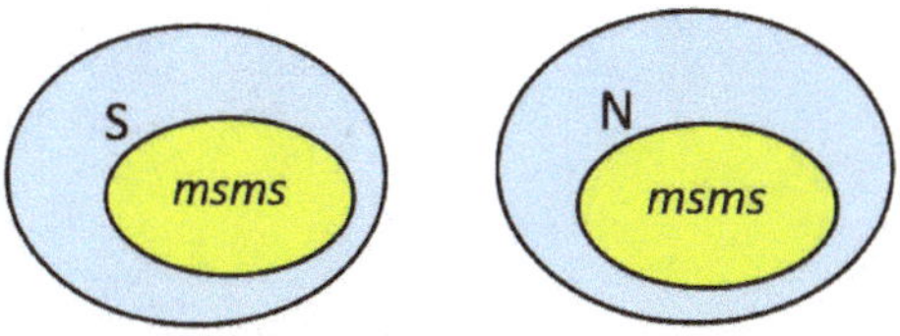

Fig. 8: Cytoplasmic male sterility system

Environment-sensitive male sterility: This is a unique male-sterility system where the expression of male sterility and fertility of the plants is controlled by environmental factors. Under this system, the male sterility gene expresses only under specific environment such as low or high temperature, short or long photoperiod, variable light intensity, different soil-borne stresses, or their specific combinations (Kaul, 1988). This situation can arise both in genetic as well as cytoplasmic nuclear male sterility systems. According to Levings *et al.* (1993) the reversal of male sterility into male fertility is influenced by cytoplasmic rather than nuclear genetic factors. The environment-sensitive male sterile line was first reported by Shi (1981) in rice; and later Yuan (1987) proposed its use in hybrid rice breeding program. Since it eliminates the requirement of maintainer 'B' line, this hybrid system, is popularly called as 'two-parent hybrid' breeding. At present this male sterility system is being used in China commercially. In 1994, the hybrids based on this system of male sterility covered over 30,000 ha areas and yields as high as 8-9 t/ha (Lu *et al.*, 1994).

In pigeonpea, an environmental sensitive male sterility system was bred from a cross involving *C sericeus* and a cultivated genotype (Saxena, 2014). The process of conversion of male sterile genotypes into male fertile with change in the temperatures is summarized in Table 4.To make practical use of this genetic material in hybrid breeding program, the seed production sites with strict temperature regimes and least fluctuations need to be identified (Fig 9). This would allow complete expression of the gene (s) responsible for this unique behavior of the genotypes. For multiplication of female parent, the maximum safe mean temperature during crop growth, particularly reproductive phase, should not exceed 20^0C. This temperature bar will maintain pollen fertility status of the plants and allow production of fertile flowers and normal pod set. In case the temperature at such sites shoots up for a short period due to some sudden changes in weather conditions, it will also not affect seed quality of the female parent. This is because if some flowers revert back to male-sterility and get pollinated by neighboring fertile flowers, then the seeds harvested from such pods will also produce male-sterile plants in the subsequent generation under warmer rainy season. The seed thus produced will remain genetically pure in spite of minor temperature fluctuations.

Table 4: Segregation for male sterility and fertility in thermo-sensitive male-sterile selections at Patancheru

Selection	September		November		March	
	Sterile plants	Fertile plants	Sterile plants	Fertile plants	Sterile plants	Fertile plants
PTSL 1	37	0	0	37	37	0
PTSL -2	32	0	0	32	32	0
PTSL -3	27	0	0	27	25	0
PTSL -5	23	0	0	22	21	0

Environment 1 (Temperature <24⁰C)

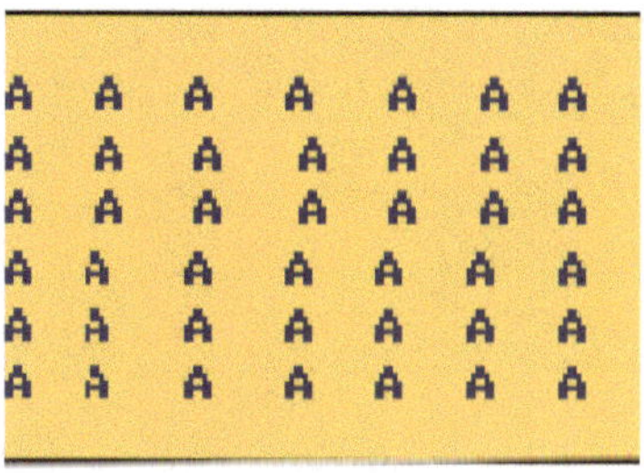

Fertile (A line multiplication) B line not required

Environment 2 (Temperature >26⁰C)

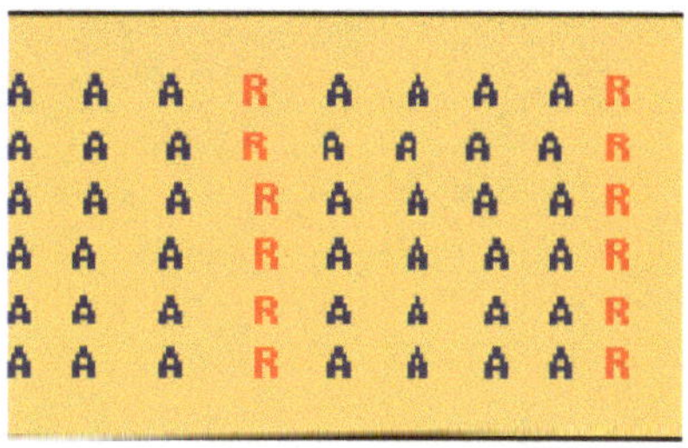

A line Sterile A x R line (Hybrid seed production)

Fig. 9: Temperature sensitive male sterility

The hybrid seed production involving temperature-sensitive pigeonpea male-sterile lines should generally be done in rainy season when the temperatures are well over 26⁰C and to avoid low temperatures, the high altitude locations should not be selected. Rouging of the female parent would be essential to eliminate any fertile plant arising due to short spells of temperature alterations. The planting date experiment at Patancheru showed that in early maturity pigeonpea genotypes both seed production of female parent and hybrid is possible at a single location. In this system the hybrid seed production plot needs to be sown during early rainy season (June); it will flower in about 60-70 days and all the flowers will be male-sterile due to prevailing high temperatures. The cross-pollinated pods that would set on these plants can be harvested in another 45-50 days. However, for optimizing yields suitable agronomic package needs to be developed. The multiplication of female parent can be taken up in the other isolation and its sowing should be done in the month of September. In this isolation all the flowers will be fertile due to low temperature and a good harvest of female parent can be taken without using B-line and pollinating insects.

3

Hybrid Seed Production

3.1. Natural cross-pollination

Breeding procedures in any plant species of economic importance are guided by its reproductive system particularly, pollination mechanism. In the cereal crops such as maize, pearl millet etc., where cross-pollination is facilitated by wind, the breeders have developed schemes for commercial exploitation of dominance and epistatic genetic variation through hybrid and population breeding. In contrast in self-pollinated crops, the efforts are always made to exploit additive genetic variation primarily through pedigree and mass selection.

In pulses, self-pollination is a rule and pigeonpea is no exception. However under field conditions, occurrence of hybrid plants in the populations is not uncommon; and it happens due to some degree of natural out-crossing that took place in the preceding generation. This type of partial natural out-crossing in pigeonpea is affected by a variety of insects (Fig 10) which forage on its large flowers to collect the nectar. These minor incidences of out-crossing are the primary source of varietal deterioration. Recently, the pigeonpea breeders converted this constraint into an opportunity for genetic enhancement of yield through exploitation of hybrid vigor and developing commercial hybrids.

Pollinating insects: Pathak (1970) was the first to identify some pollinating insects in pigeonpea and these were *Megachile bicolor* and *Apis florae*. In a detailed study conducted at Patancheru (Andhra Pradesh) Williams (1977) identified 48 insect species foraging on pigeonpea, but the common foraging insects were *Apis dorsata* and *Megachile* spp. (Saxena *et al.*,1990).

Apis mellifera *Apis dorsata*

Apis florea *Megachile flavipes*

Fig. 10: Insect pollinators of pigeonpea

Extent of cross-pollination: Experiments conducted at Patancheru under controlled conditions excluded the possibility of any wind pollination in pigeonpea (Kumar and Saxena, 2001). Hence, the entire event of cross-pollination can be attributed to the foraging insects. Therefore, the extent of cross-pollination at any given location is determined determined by the type of insects and their number. Besides this, some key factors (*e.g.* wind direction and its speed), the density of insects and their movement also influence the extent of natural out crossing at a particular location. The areas with bushy habitat and some water resource which support insect harboring and thereby extend foraging activity in pigeonpea fields (R.V. Kumar, Personal communication). The fresh large yellow/ red flowers attract the pollinating insects and during the process of nectar, about 5000-100,000 pollen grains get attached to the body of the insects. These insects when visit other flowers cause cross pollination (Williams, 1977). Dalvi and Saxena (2009) demonstrated that stigma receptivity in pigeonpea flowers extends from 2-days prior to and one day after flower opening. It is believed that the extended stigma receptivity also helps in cross pollination in pigeonpea. Byth *et al.* (1982) reported that even the morphology of flower, especially the orientation of petals, also play a significant role in determining natural out-crossing in pigeonpea. As a consequence of various factors described above, the natural out-crossing in pigeonpea is inconsistent over the locations (Table 5).

Table 5: *Natural out-crossing (%) in pigeonpea at different locations

Location	Range	Mean	References
India			
Pusa	2.3-12.0	6.7	Howard *et al,* 1919
Nagpur	3.0-48.0		Mahta & Dave, 1931
Niphad	11.6-20.8	16.0	Kadam *et al,* 1945
Coimbatore		13.7	Veerswamy *et al,* 1973
Varanasi	10.3-41.4		Bhatia *et al,* 1981
Badnapur	0.0-8.0		Bhatia *et al,* 1981
Hyderabad	4.0-26.0		Saxena *et al,* 1987
Kenya	17.7-45.9		Onim *et al,*1981
Hawai	5.9-30.0	15.9	Wilsie & Takahashi,1934
Australia	2.0-40.0		Byth *et al,*1982
Uganda	8.0-22.0		Khan, 1973

*For references and other details see Saxena *et. al.* (1990)

3.2. Important features of isolation plots

A universal isolation distance for pigeonpea seed production is difficult to recommend due to various physical factors described in section 3.1; and these cannot be standardized like other crops. Hence the local breeder/seed producer needs to use his/her judgment using the past experience. Based on the information collected from different locations over the years, ICRISAT recommends an isolation distance of 500 m for certified seed production of hybrids and 1000 m for Breeder and Foundation seed production of hybrid parents. The other important aspect in selecting the sites for the seed production of hybrid and female parent is that the isolation should be conducive for harboring honey bees and for this, the natural habitat should be infested with bushes or honey bee harboring fruit and other trees. In such sites, the bee activity is extended and pod set is high. Such sites when located near temporary or permanent water bodies can further enhance setting of crossed pods on the male sterile plants and yields over 2500 kg/ha have been harvested repeatedly (RV Kumar, personal communication).

3.3. Planting design

In addition to site selection as described above, the planting design or field plot techniques are also important for optimizing hybrid yields. The main focus should retain the pollen parent plants in flowering state as long as possible. This will extend the visits of pollinating insects to the male fertile flowers and maximize the cross pollination. To avoid water stagnation, it is always recommended to plant the seed production plots on ridges, about 100 cm apart. The intra-row spacing should be kept at 30 cm. As a thumb rule, the net seed production plot containing 4 female:1 male rows should be surrounded by a 3-4 meter long belt of pollen parent, preferably 2-3 weeks later than the net plot (Fig 11) If the population of pollinating insects is low then the materials in the protective outer belt can also be split into two sowing dates. This technique works well in small seed production plots of 1.0-1.5 acres. In the large seed production plots, it is recommended that the entire plot be divided into 2-3 sub plots, depending on the total plot size. Each sub plot should be planted with 4 female and 1 male combinations. The sowing of each sub-plot, containing male and female rows, should be staggered by two weeks (Fig 12).

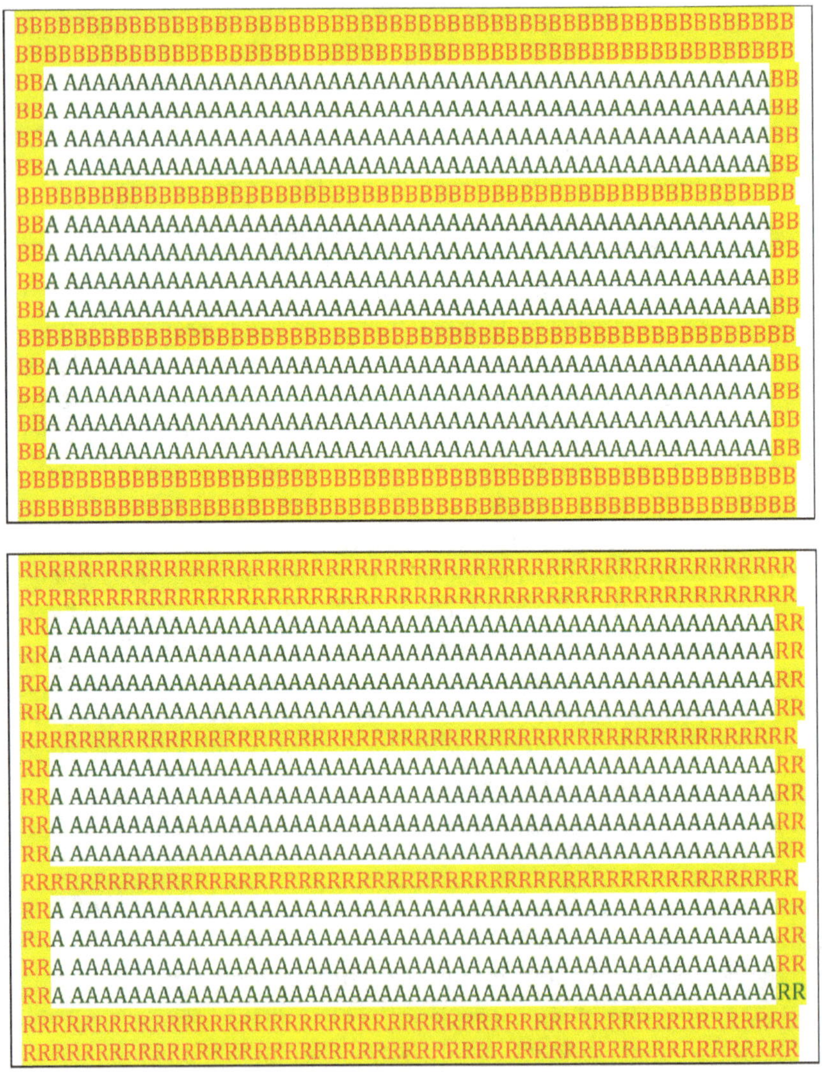

Fig. 11: 4F:1M surrounded by 2 rows of male lines

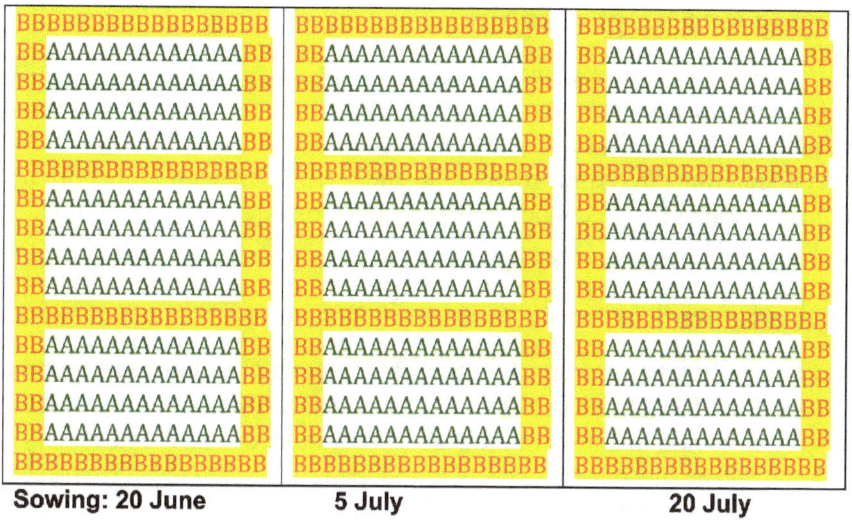

Fig. 12: Staggered sowing of sub plots by 2 weeks interval

3.4 Crop management

In pigeonpea a large variation exists for maturity (85-300 days) and plant type (determinate and non-determinate). These factors lead to some markedly different phenology and canopy types. Therefore, uniform agronomic practices cannot be recommended for optimizing yield in different agro-ecologies. Similarly, determinate types have small plants while the long-duration types produce about 8-10 times more biomass, horizontally as well as vertically. Since pigeonpea plant is sensitive to photoperiod and temperature, the growth rate and biomass production of the plants are very different at different latitudes and planting dates. For example, a variety like UPAS 120 will flower in about 70 days and attain a height of 1.5 m at Patancheru (17° N); while at Hisar (29° N) on the same date of planting , it will flower in about 90-100 days and grow over 2 m. Hence, for optimizing yield environment-specific agronomic package need to be worked out. However, some general crop production guidelines for pigeonpea are discussed here.

Climate

Pigeonpea grows well in warm tropical and subtropical climate. The crop prefers a fairly moist and warm climate during the period of its vegetative growth. During the flowering and ripening stages of its growth, it requires bright sunny weather for the setting of pods. It is highly susceptible to frost at the time of flowering. Cloudy weather and excessive rainfall at flowering time damage the crop to great extent.

Soil and land preparation

Soil: Pigeonpea may be grown well, on a wide range of soils varying from sandy loams to clay loams. It does best on fertile and well drained loamy soils. The saline-alkaline and waterlogged soil unfit for its cultivation, as they adversely affect nodulation. Well drained, alluvial and loamy soils with proper drainage are most suitable for pigeonpea cultivation. It grows well in well drained red clay loam soil too.

Land preparation

Land preparation for pigeonpea requires at least one ploughing during the dry season followed by 2 or 3 harrowing. The "summer" plowing helps in minimizing the weed flora and to conserve moisture. Well-drained soils are necessary for good root and nodule development. Contour beds or a ridge-and-furrow system are useful in preventing water logging by draining excess surface water, and in preventing soil erosion

Sowing time: An appropriate sowing time of the crop is end of June to first week of July. The delayed sowing may lead to frost damage in January causing drastic yield reduction.

Seed rate: Usually farmers put very high seed rate of upto 25 kg/ha. However, the recommended seed rate is

Early maturity: 15 kg/ha

Medium-maturity: 12 kg/ha

Hybrid: 8 kg/ha

Spacing: (Row to row x Plant to plant)

Early maturity: 60 x 20 cm

Medium maturity: 75 x 25 cm

Hybrid: 75 x 30 cm (Rainfed)/ 100 x 30 cm (Irrigated)

Seed treatment

Bio-cultures and fungicides: Before sowing, seeds should be treated with 2 g Thiram + 1 g Carbendazim fungicides per kg of seed. Thereafter, it should be treated with 5 g PSB (Phosphate solublizing bacteria) + 5 g Pigeonpea rhizobium culture per kg of seed.

Sowing method

Sowing should be done either through Bullock drawn Tifan/Mahakaal Tifan or tractor drawn seed drill with appropriate recommended spacing.

Ridge sowing of pigeonpea allows the excess water to drain and favours better plant growth. It also protects the crop from *Phytophthora* disease incidence at seedling stage.

Nutrient management

Major nutrients: For exploitation of potentials of varieties, recommended dose 20 N: 50 P: 20 K: 20S kg/ha

Micro nutrients: Application of Zn @ 25 kg/ha once in three year

Manures and fertilizer requirement: 10 t FYM/ha or 5t FYM/ha +100 kg DAP/ha +33kg Mop/ha+200 kg Gypsum/ha.

Methods and time of manures and fertilization application

Being legume crop, the manures and fertilizers are applied as basal dose to pigeonpea. Well decomposed compost and fertilizer if given in the form of single super phosphate should be spread before last harrowing. Sulpur in the form of Gypsum should also be broadcasted before last harrowing. DAP and Murate of potash should be given at the time of sowing below the seed zone with the help of fertilizer cum seed drill.

Water management

Pigeonpea is mostly grown as rainfed crop in Madhya Pradesh and other states of the country. However, wherever, irrigation facilities are available, only one irrigation at the time of flowering is sufficient for better pod filling. No irrigation should be given after complete pod formation, as it may lead to further flush of flowers and restrict the pod filling of earlier flush, causing disrupted pod filling.

Weed management

Pre-emergence: Pendamethylene @ 1 kg a.i. /ha (3.33 Lit/ha) just after sowing

Post-emergence: Imazethapyr @ 50 g a.i./ha (500 ml/ha) after 30-35 days of sowing is recommended with one hand weeding.

Insect control: Insects like pod borer *(Helicoverpa armigera* and *Maruca vitrata*) ; pod sucking bugs (*Clavigrella* spp); and pod fly (*Melanagromyza obtusa*) are major pigeon pea insect pest may cause serious reduction in yield and grain quality.sometimes, a total crop loss may occur . The insecticide such as dimethoate 30EC@1lit/ha ; fluebendiamide 48EC@ 240ml/ha; spinosad 45 EC @ 240ml/ha have been found effective against the pests. The first spray is recommended at 50% flowering stage and should be repeated after 15 days, if required .In case of high volume sprayer are used then 800 lit/ha spray solution recommended. Here is a word of caution –although imidacloprid,thiacloprid (neonicotinoids) are effective against the insect pests but this group of insecticide is harmfull for the pollinators specially bees and wasps, hense their use on pigeonpea must be discouraged. In storage, brushes are considered as important pests as they inflict direct losses to the grain/seeds and more over, there is no compensation for damage. Hence cleaned and well dried grains should be stored in sealed containers with fumigants like alluminium phosphide @3g/bag (40kg bag) . If seeds are to be stored, mixing seed with fine ash or attapulgite based clay dust@ 1:10 can help to minimize storage insect problems. The fumigants and inert materials do not affects the seed viability/ germination.

The crop should be harvested by cutting the plants at ground level at 75% pod maturity. The threshing can be done as per the local practices. Cleaned and well dried grains should be stored in weevil-free containers but fumigants can also be used if found essential to kill the storage pests.

3.5 Recognized seed classes and their production

Seed production of A, B & R Lines

In India there are five approved seed classes. The main features of the seed classes and their seed production procedures, as applicable to the hybrid pigeonpea, are briefly described below:

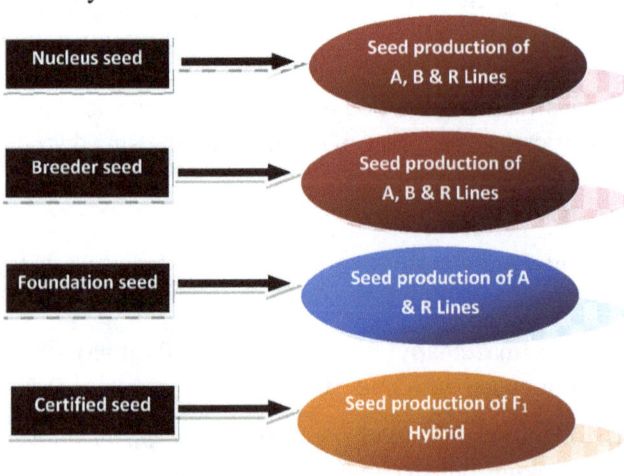

7Fig. 13: Seed multiplication chain in hybrid seed production

Nucleus seed production: The nucleus seed of hybrid parents should be produced with the highest standards of genetic purity. For multiplying nucleus seed of male sterile (A-) line, the available nucleus seed of A- and B- lines should be acquired from a reliable source. The seed lots should be examined for any physical impurity before undertaking their multiplication. The sowing of these seed lots should be taken under an insect proof net in adjacent plots and all the recommended cultural practices should be followed. At flowering, each and every plant of A-line should be examined for any trace of pollen grain; and any suspicious plant should be removed. Similarly, each plant of B-line should be examined for the known morphological traits and roguing should be done to remove any off type. The seed multiplication should be organized by hand pollinating single plants of A-line with single plants of B-line. The hybridization should continue till the required number of pods is set in the whole plot. The hand harvesting and threshing of the crossed pods should be done with a lot of care. The seed thus harvested should be labeled, dried to about 9-10% moisture, and stored in a cool dry place.

The maintainers (B-line) and restorers (R-line) are male fertile and hence their nucleus seed production is rather easy. This grade of seed should also be multiplied under insect proof nets. In this case also the individual plants should be examined for their known morphological traits and intensive roguing standards should be adopted. After careful harvesting, threshing and labeling, the seed can be stored.

Breeder seed production: In this category of seed production also, very high standards of purity should be maintained. This seed is multiplied in isolation and a minimum of 1000 meter of isolation is recommended. For the production of the seed of male sterile (A-) line, nucleus seed of A- and B- lines should be sown in isolation in the ratio of 2 female: 1 male row. At flowering intensive roguing of A-line should be taken up by examining each plant for the absence of pollen grains and any suspected plant should be removed. The population should also be examined for purity using key morphological traits of the male (B-) and female (A-) parents. At maturity, the male rows should be harvested first by cutting the plants from ground level and removing them out of field.

The breeder seed production of B- and R- lines is under taken in different isolations with a minimum distance of 1000 meter. It is recommended that after intensive rouging at different stages about 100 representative single plants should be harvested separately from the center of the isolation. These should be threshed, cleaned, labeled, dried and stored carefully for use in future seed production programs.

Foundation seed production: This lot is a pre-requisite of certified grade of seed; therefore relatively large quantities of the parental seed need to be produced (Fig 12). The seed of all the three parents is produced in different isolations located at least 500 meter apart. For seed multiplication of A-line, the female to male ratio of 4:1 is generally recommended. But this can be reduced to 3:1 if the insect pollinators are limiting. Examination of each plant in the female rows should be done for the expression of 100% male sterility and any suspected individual should be removed. At maturity, the male rows should be harvested first and removed from the field. The field operations described earlier for breeder seed production should be adopted here also.

Certified hybrid seed production: The production of hybrid seed for distribution to farmers is classified under 'Certified Seed' grade. Under this grade large quantities of seed are produced by various seed organizations and institutions. The seed production of hybrid is undertaken in isolation, at least 500 meter away from other pigeonpea crop. Generally a combination of 4 female: 1 male row ratio gives good results; but this ratio can be altered upwards or downwards, depending on the density of pollinating insects in the vicinity; and this knowledge is generated through experience of the seed growers. To remove off-type plants, at least two rounds of rouging are essential, one before flowering and another during flowering. At maturity the male rows should be cut by hand and removed from the field to avoid the chances of seed mixture. As described above the seed should well dried and treated before storing.

Truthfully labeled hybrid seed production: This seed grade was allowed by the Government to meet the massive seed requirement and inability of seed organizations to meet the domestic demand. In this type of seed production various registered organizations such as farmers' groups, NGOs etc. can produce and market the hybrid seed and all the responsibilities of seed quality etc. lie with the producer. The field operations are the same as described above.

3.6 Productivity of hybrid seed

The new (modified) seed production technology has been successfully tested in a number of locations (Table 6). The mean hybrid yield across 94 locations was 1019 kg/ha. Further, it was also observed that under good crop management and site selection, the hybrid yields can be increased by a strong margin of 75-100%.

Table 6: Hybrid seed production (A x R) data (kg/ha) recorded in six states

State	Locations	Mean yield	Highest yield
Andhra Pradesh	34 (6)	998	1750
Madhya Pradesh	9 (3)	1674	2267
Gujarat	4 (2)	1179	1669
Maharashtra	5 (2)	603	1017
Odisha	40 (1)	523	1040
Karnataka	2 (2)	1138	1900

( ) number of years

3.7 Seed quality determination

Seed quality control gets very high priority in every commercial hybrid breeding program. In the short-aged crops it is easy to test the quality of hybrids and their parents by evaluating their progenies for some characteristic morphological markers controlled by 1or 2 genes. Since pigeonpea is a long duration (around 200 day) crop and raising two crops in a year is impossible. This constraint has limited the adoption ofhybrid technology by prominent seed companies. Recently, two alternatives have been invented to overcome this bottleneck.

Use of morphological trait: The technology is based on 'obcordate' leaf (Fig. 14) marker that is controlled by single recessive gene and in the progenies it is expressed within four weeks from sowing. Since there is no seed dormancy in pigeonpea, this technology can be deployed easily and the results can be obtained immediately after the harvest of the crop. The male sterile female (A-) lines with obcordate leaf marker have been developed (Saxena et al., 2011b) and all the known fertility restorers have normal (lanceolate) leaves with dominant gene control. In such combinations all the hybrid plants will have normal leaves. Since this trait is not influenced by environment, the quality testing can be done in large number of samples in a short time and with least expenses (Fig 15).

Fig 14: Obcordate leaf type in pigeonpea and its incorporation as morphological marker

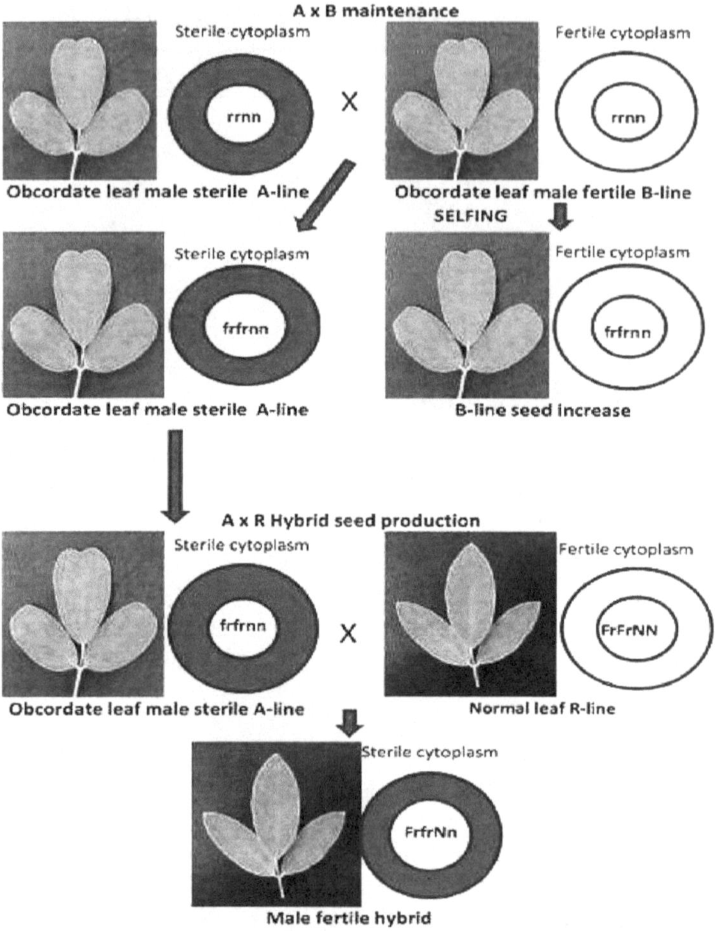

Fig. 15: Use of morphological trait (Obcordate leaf) for purity maintenance

Use of genomic science: This test may also be influenced by changes in certain environmental factors. In recent years simple, rapid, and cost effective seed quality testing approaches that are based on molecular markers assay have been developed (Naresh *et al.,* 2009; Bohra *et al.,* 2011; Saxena *et al.*, 2010) and a set of two simple sequence repeat (SSR) markers (Fig 14) for testing the hybridity of ICPH 2438 was developed. Subsequently, a set of 42 SSR markers was identified for purity assessment of the hybrid RVICPH 2671. In order to save time and costs, the set of 42 markers was sunk into eight groups of multiplexes. With the help of these markers, detection of off-types in the commercial hybrid seed lots can now be undertaken by both public and private seed companies. This assay can now be used for reliable assessment of seed purity within commercial seed lots of hybrids to ensure the supply of high quality seeds.

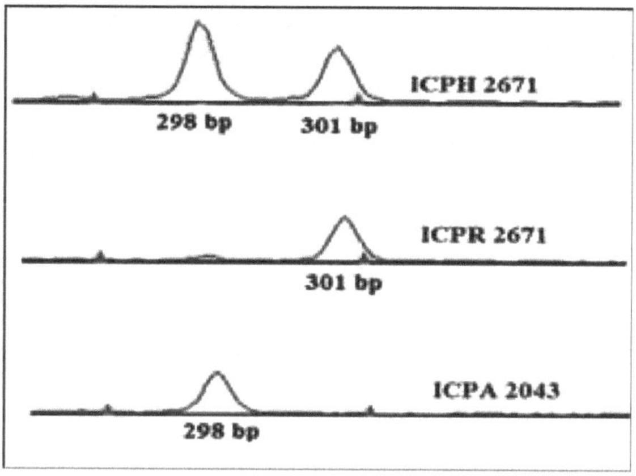

Fig. 16: A snapshot showing the hybrid purity assessment of hybrid RVICPH 2671 with the CcM 0021 marker. The parental lines ICPA 2043 (A-line) and ICPR 2671 (R-line) show 298- and 301-bp alleles, respectively, on screening with a diagnostic simple sequence repeat (SSR) marker (CcM 0021), while ICPH 2671 seed showing the presence of both alleles (298 and 301 bp) represents true hybrid (Bohra *et al.* 2011).

3.8. Cost of hybrid seed production

In order to make hybrid seed available to small holder farmers, it is essential that the hybrid seed is available at affordable prices. According to the studies conducted by Saxena *et al.* (2011) at Indore (Madhya Pradesh), the cost of producing one hectare of hybrid pigeonpea seed, excluding land cost, was Rs. 26,395 and one kilogram of hybrid seed was produced @ Rs. 18.32 (Table 7). Due to vigorous canopy of hybrid plants, the recommended seed rate for the hybrids is 50% less than pure line cultivar; this will help in the adoption of hybrids even by small farmers because it will not increase the seed input cost.

Table 7: Profitability and cost of hybrid seed production of pigeonpea hybrid RVICPH 2671 at Indore during 2007

Items	Cost (Rs./ha)
Gross expenditure	
(a) Non-labour inputs	
Field preparation	2,000
Seed	900
Fertilizer and seed treatment	3,205
Sub Total (a)	**6,105**
(b) Labour input	
Sowing (40)	3,740
Weeding & Interculture (58)	5,423
Roguing (20)	1,870
Spraying (22)	2,057
Harvesting (45)	4,208
Threshing (32)	2,992
Sub total (b)	**20,290**
Total expenditure (a+b) (Rs)	26,395
Hybrid yield harvested (kg)	1,400
Cost of producing one kg seed (Rs)	26,395 /1,400 = 18.85
(c) Cost of seed production	
Income from sale of 1400 kg hybrid seed @ Rs 60/kg	84,000
Income from sale of 800 kg R-line seed @ Rs 15/kg	12,000
Income from sale of total fuel wood	400
Total	**96,400**
Net Profit (Rs / ha)	**96,400 – 26,395 =70,005**

4

Breeding New Hybrids

4.1. Flower and flowering in pigeonpea

Growth habit: Primarily, there are two major growth types in pigeonpea. These are determinate and non-determinate. The determinate growth in the pigeonpea plants is characterized by flowers in bunches at the top of the canopy (Fig 17a). In this case the epical growing points are transformed in to reproductive buds and bunches of flowers appear on the top of the canopy. Such cultivars are not popular as the management of insects such as *Helicoverpa* and *Maruca* becomes very difficult. Most of the cultivated types belong to another growth category called as non-determinate. In such genotypes the apical growing points remain vegetative and the plants keep growing under favorable conditions and it results in tall canopy with many spreading or compact branches. The inflorescence in this case emerges from the axil of secondary and tertiary branches (Fig 17b); and hang sideways. These types show less susceptibility to insects due to non-clustering nature of flowers.

Induction of flowering: Pigeonpea is a typical short day plant and induction of flowering is controlled by photoperiod, temperature, or their interaction. It has been observed that in pigeonpea earliness is related to photo-insensitivity; but none of the genotype so far has been found to be truly insensitive (Saxena, 2008). The growth phases in this crop have not been properly defined so far and information on the length of inductive phase is also not known.

Flower structure: Pigeonpea has a typical legume flower with 10 (9+1) anthers enclosed in keel petal (Fig 18). The anthers are arranged in two whorls, the upper whorl has bigger anther lobes as compared to the lower whorl. It is believed that the upper anthers participate in cross-pollination and the lower one in self-pollination.

Stigma receptivity: For hybridization the duration of stigma receptivity plays an important role as it will provide extended periods for pollination and successful fertilization. Dalvi and Saxena (2009) made an elaborate study on this aspect in pigeonpea. They concluded that stigma of pigeonpea becomes receptive for fertilization three days before flower opening and remains in the same condition two days after flower opening. This means successful pollinations can be made to produce hybrid seed by hand pollination for a period of 5-6 days (Fig 20 a & b).

Fig 17 a: Determinate

Fig 17 b: Non-determinate

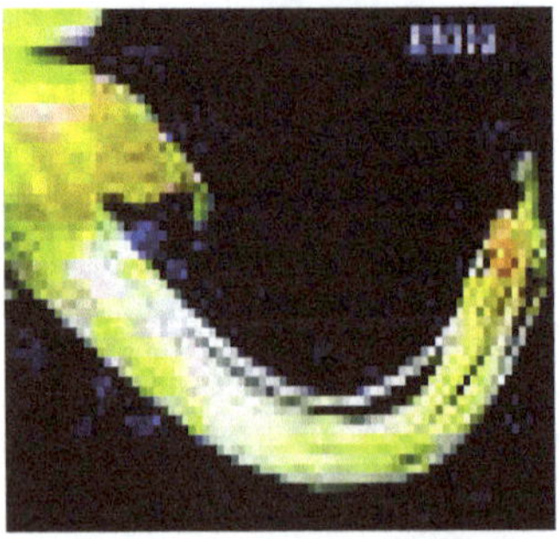

Fig 18 a: Di-adelphous condition of stamens

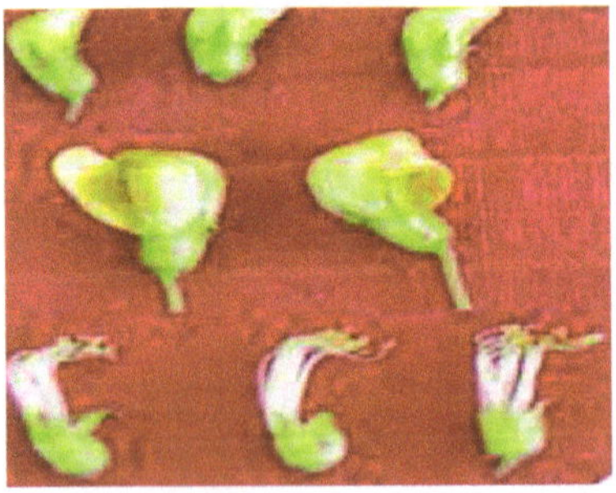

Fig18 b: Stage of flower for emasculation

Fig 18 c: Standard, petal & Keel covering sigma & anthers

Regeneration of flowering: Botanically, pigeonpea is classified as a short-lived perennial species; and it can easily survive 3-5 years and its wild relatives survive even for longer periods, provided it is not attacked by any lethal external factor. It is a survival mechanism that allows production of more than one reproductive flushes on the same plants within or over years. Since early maturing types are relatively more insensitive to changes in photoperiods, these type continue to flower for a longer period of the year and therefore allows hybridization for a longer season. On the contrary, the late maturing types being highly sensitive to photo-period and therefore their flowering duration is restricted. These types enter into vegetative phase in the second year of growth and flowering in these types will be induced only at the onset of short days. The regenerated reproductive growth can also be used for multiplying some important genotypes. Saxena *et al.* (1976) exploited this phenomenon for completing a huge 28 x 28 diallele involving 378 crosses in one season. This program involved early diverse flowering (60-150 days) parental lines and hybridization between early and very late types was possible by using the regenerated growth of the early flowering lines; and this trait of pigeonpea helped in making the gigantic hybridization program a success.

4.2. Genetic diversity and selection of parents

Breeders are aware of the importance of genetic diversity in crop improvement, particularly hybrid breeding(Fig 19 a). The hybrid, once perfected, will provide instant yield advantage through the gene actions such as dominance, over-dominance and epistasis through the manifestation of heterozygosity at various favorable loci. Richey (1922) demonstrated for the first time the positive relationship between genetic diversity and hybrid vigor in maize. A number of studies on the relationship of genetic diversity and heterosis led to the concept of 'heterotic groups', where the parental lines of diverse groups yielded good results. This concept has been used extensively in maize (Jelena *et al.,* 2007) and rice (Yuan, 1987) hybrid breeding programs and dividends in terms of increased yields were encouraging .

In pigeonpea, the first such effort was made by Saxena and Sawargaonkar (2014). They demonstrated that the hybrids between genetically diverse heterotic groups produced >100% yield advantage; and on the contrary the crosses between genetically related heterotic groups, the large yield advantage was missing (Fig19 b).Further,they recommended that the problems associated with morphological data and g x e interaction, information on genomic diversity can be very useful. Using the technology of heterotic group, the cost of hybrid breeding can be restricted; and the classification of germplasm into different heterotic groups can be done on the basis of F_1 performance of crosses, origin,

Fig 19a: Genetic diversity in pigeonpea accessions using molecular markers (Rachit Saxena *et al*, 2014)

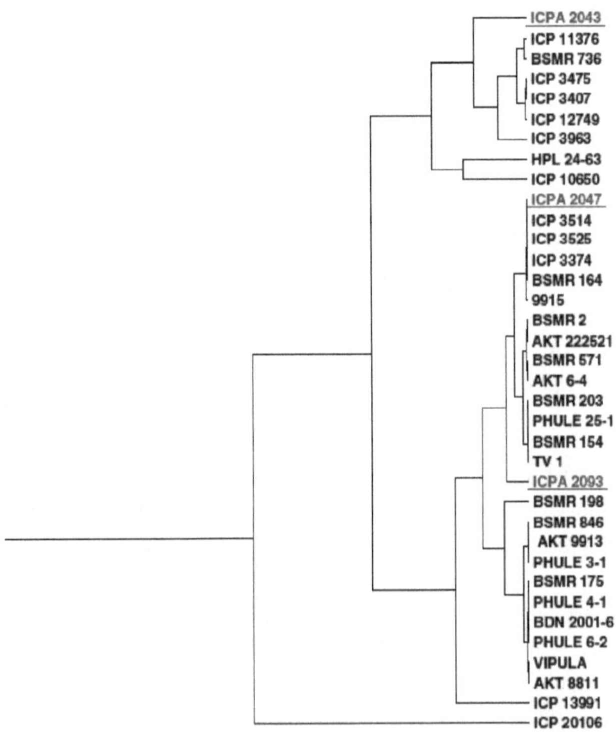

Fig 19 b: Heterotic groups in Pigeonpea (Saxena and Sawargaokar, 2014)

and genetic diversity of the parents. This will avoid the expenses involved in carrying the load of some unproductive cross combinations in the breeding program.

4.3 Hybridization

Hybridization using manual resources is a serious business since it requires pollination skill and management of plant materials. Crosses between the two designated parents should begin with close observations on uniformity and purity of the parental lines and any off-type should be removed. The selection of flower buds of appropriate age for crossing and pollen source are important. A thumb rule for selecting such buds for crossing is when the lengths of its calyx and corolla are approximately equal. The selection of buds as pollen source is also important so that it has viable pollen grains (Fig 18). This can be done by selecting fully grown unopened buds with bright color. In this case the ratio between calyx and corolla length should be approximately 1:2. Each pollinated bud should be tagged and at maturity care should be taken in harvesting, drying, threshing and storing to avoid seed mixtures.

Fig. 20: Emasculation and hand pollination for hybridization

4.4 New male and female parents

Frequency of fertility restorers and maintainers

A fertility restorer line is defined as a genotype which restores the fertility of the progeny. Hanson and Bentolila (2004) and Wang *et al.* (2006) reported that CMS is a function of certain unusual open reading frames coding for a polypeptide chain. Male fertility in such genotypes can be restored by specific nuclear genes which encode fertility restorer genes through the production of pentatricopeptide. Restoration of fertile pollen production on the male-sterility based hybrid plants is the key factor in exploiting hybrid vigor in sexually reproducing crop species. This generally happens when dominant fertility restoring nuclear gene(s) present in the male-parent are transmitted to the hybrid plants. Such genes repair the damage caused by mitochondrial DNA aberrations in the male-sterile plants. Recent inheritance studies by Saxena *et al.* (2011a) and Sawargaonkar *et al.* (2012) revealed that restoration of pollen fertility in A_4 CMS system of pigeonpea was controlled by either single dominant or two duplicate dominant genes. Saxena *et al.*(2011a) further reported that for stability of fertility restoration across diverse environments, presence of both the dominant genes was essential.

In pigeonpea, the fertility restoring genes are sporophytic in nature (Dalvi *et al.,* 2008) and hence both homozygote and heterozygote hybrid plants produce fully fertile pollen grains. According to Singh and Gopalkrishnan (2013), the frequency of fertility restorers among the cultivated types for an alloplasmic CMS system is generally low due to negative association of genetic diversity of the parents with the fertility restoration of F_1 hybrid. On the contrary, in the present alloplasmic CMS system, the frequency of fertility restorers was reasonably high (35.7%) and this situation may arise due to genetic closeness of *C. cajanifolius* with *C. cajan* (De, 1974; van der Maesen, 1990; Mallikarjuna *et al,* 2012). Therefore, it can be assumed that the mitochondrial defects in *C. cajanifolius* caused by insertion of *C. cajan* genome were not of serious nature and these can be repaired easily by fertility restoring genes present in the primary gene pool of genus *Cajanus*.

In the early maturing group 35 fertility restorers were identified (Table 8). The flowering and maturity periods among the early maturing restorers varied from 50 to 85 and 101 to 141 days, respectively (Table 9). ICP 3868, ICPL 90012, and ICPL 90051 had seed size of 10g /100 seeds or more (Table 9). Like maintainers among the restorers also, resistance to diseases in this maturity group was limiting and only ICPL 89032 had resistance to fusarium wilt, while ICP 14057, ICPL 5, ICPL 92043, and 93107 were resistant to sterility mosaic virus. Six restorers viz., ICPLs 89, 90030, 90036, 90048, 93103, and 93107 had white seeds (Table 10). Medium maturity group is very important from adaptation point of view (Saxena, 2008) and hence, a large number of crosses were attempted and 113 fertility restorers were identified (Table 8). These represented a fairly good genetic variation with respect to key plant characters (Table 9). In this group 72 restorers were resistant to both wilt and sterility mosaic diseases. The variation for maturity was from 138 to 200 days. Sixty-nine testers had seed size of ≥10 g/100 seeds. In the late maturing group 31 restorers were identified (Table 8). These included 11 from Africa and 16 from India. In this group 18 testers were resistant to sterility mosaic virus; while only five exhibited resistance to *fusarium* wilt. ICPL 20103, MA 16, ICPL 20120, ICP 11376, ICP 13092, and ICP 14282 were found resistant to both the diseases. Plant maturity in this material ranged from 186 to 241 days. Two testers ICP 13379 and ICP 8051 had seed size of 18 g/100 seeds; while seven recorded seed size of >15 g/100 seeds.

Saxena *et al.* (2014) evaluated 502 testers and reported that 179 (35.7%) restored male fertility in the hybrid plants. In contrast, the frequency of male-sterility maintainers was quite low and only 25 (5.0%) lines maintained male-sterility. The remaining 298 (59.4%) crosses had variable proportions of male-sterile and fertile plants (Table 8).This situation may arise due to heterogeneity for fertility restoring genes within testers.

Out of 179 restorers identified, 35 were of early maturing group, 113 were of medium maturity, and 31 represented late maturity group. Similarly, out of 25 maintainers, eight represented early, 15 medium, and two late maturities. The high frequency of fertility restorers may be due to genetic closeness of *C. cajanifolius* with *C. cajan.* The plant and grain characteristics of 35 elite restorers showed a considerable variation for important agronomic traits and this provides options to breeders for selecting desired hybrid parents (Table 10). The flowering and days to maturity among the restorers ranged from 107 to 154 and 172 to 204 days, respectively. The restorers ICPL 20098 and ICPL 20104 had the largest seed size. Twenty one restorers were found resistant to both wilt and sterility mosaic diseases. Seven inter-specific derivatives involving *C. acutifolius, C. platycarpus, C. scarabaeoides,* and *C. lineatus* restored pollen fertility of A_4 CMS system and provided additional variability for any hybrid breeding program. These fertility restorers exhibited a high levels of pollen production ability over diverse environments and years (Table11). To breed new restorer lines, crosses among selected diverse restorers can be made to identify desirable genotypes with respect to different consumer preferred traits. Further, based on the genetic diversity, combining ability and *per se* performance a set of heterotic groups be developed for use in hybrid breeding programs.

Stability of fertility restoration

A total of 35 restorers were used to study stability of pollen fertility in hybrid combinations at diverse locations in different years. Of these, 20 were evaluated at 10 environments for three years (Table 11). Their mean pollen fertility ranged from 88-99 %. The remaining 15 hybrids were evaluated in seven environments for two years and their pollen fertility ranged from 85.5 to 100%. The results of these multi-location trials showed that the testers were highly stable in their ability to restore fertility across diverse environments. The plant and grain characteristics of 35 elite restorers (Table 10) showed a considerable variation for important agronomic traits and this provides options to breeders for selecting desired hybrid parents. The flowering and days to maturity among the restorers ranged from 107 to 154 and 172 to 204 days, respectively. The restorers ICPL 20098 and ICPL 20104 had the largest seed size. Twenty one restorers were found resistant to both wilt and sterility mosaic diseases.

Maintainers

A maintainer line is defined as a genotype which maintains the fertility of the male-sterile lines. About 25 maintainers were identified as maintainers. These included eight early, two late and 15 medium maturing types. The plant and grain characteristics of the maintainers are given in Table 12. Plant maturity among the early types ranged between 95 and 127 days; and only ICP 16172 and ICP 14425, had large seeds. There was no resistance to *fusarium* wilt in this group and only ICP 14857 and ICP 14849 exhibited resistance to sterility mosaic virus. All the early maturing maintainers had brown seeds. In the medium maturing group, resistance to *fusarium* wilt and sterility mosaic diseases is of prime importance (Reddy *et al.*, 1990). The data recorded in the disease screening nursery revealed that 10 out of 15 maintainers had resistance to both the diseases (Table 12). Four testers (ICPLs 20286, 20287, 20288, and 96053) had white seeds and ICPL 20099 had the largest seeds. ICPL 118 was determinate in growth habit, while the rest were non-determinate. One of the medium maturing maintainers, ICP 5529 had a special leaf marker, identified as ''obcordate" (Fig. 14). This trait is controlled by a pair of recessive alleles (Saxena *et al.,* 2011b) and it is expressed within 25-30 days from sowing. This trait can be used to maintain genetic purity of male-sterile lines and hybrids with minimum efforts. In the late maturity group, only two male-sterility maintainers were identified. ICP 14085 matured in 193 days and had good seed size. It was tolerant to wilt and sterility mosaic diseases, each recording 20% incidence. The other maintainer in this group was ICPL 20092. It is a white seeded line with tolerance to wilt and resistance to sterility mosaic virus. These two maintainers can easily be purified for disease resistance with careful selection of resistant male-sterile plants in the disease-sick nursery and backcrossing them with resistant single plants of the recurrent parents.The data recorded in the disease screening nursery revealed that 10 out of 15 maintainers had resistance to both the diseases. The list of male-sterility maintainers and their important traits are given in Table 12.

Table 8: Frequency of fertility restorers and male-sterility maintainers of A_4 cytoplasm in different maturity groups

Group	Early	Medium	Late	Total
Maintainers	8	15	2	25 (5.0)
Restorers	35	113	31	179 (35.7)
Segregating	65	205	28	298 (59.4)
Total	108	333	61	502

( ) Percent

Table 9: Variation for important traits among fertility restorers of early, medium, and late maturity groups recorded at Patancheru

Trait	Early(n= 35)	Medium(n= 113)	Late(n= 31)
Days to flower	50 – 85	90 - 130	131 – 158
Days to mature	101 – 141	138 - 200	186 – 241
Plant height (cm)	70 – 165	90 - 228	135 – 260
100-seed weight (g)	6.2 – 12.1	6.8 – 17.3	7.7 – 18.1
Wilt (%)	52 – 100	0 – 100	0 – 100
Sterility mosaic (%)	3 – 67	0 – 100	0 – 100

Table 10: List of elite medium maturing fertility restorers and their important traits

S.No.	Genotype	Days to flower	Days to mature	Plant height (cm)	100-seed weight (g)	Wilt %	Sterility mosaic%	Seed colour
1.	ICPL 87119	122	172	228	10.6	0	0	Brown
2.	ICPL 20093	123	180	190	12.6	8	0	Brown
3.	ICPL 20096	120	176	155	10.9	0	0	Brown
4.	ICPL 20098	122	177	180	13.0	0	0	White
5.	ICPL 20104	120	178	190	12.9	7	0	Brown
6.	ICPL 20106	122	179	185	12.5	9	0	White
7.	ICPL 20107	119	173	162	8.6	20	0	Brown
8.	ICPL 20108	119	177	192	10.9	0	0	White
9.	ICPL 20111	122	181	192	10.4	15	0	Brown
10.	ICPL 20112	120	178	188	8.8	14	0	White
11.	ICPL 20116	116	175	148	10.8	0	0	Brown
12.	ICPL 20120	139	199	185	9.8	3	1	Brown
13.	ICPL 20123	121	182	170	11.6	0	0	Brown
14.	ICPL 20125	120	181	170	10.4	20	20	Brown
15.	ICPL 20127	125	184	162	10.4	85	15	Brown
16.	ICPL 20128	122	181	198	11.3	0	0	Brown
17.	ICPL 20129	129	192	195	12.1	13	0	Brown
18.	ICPL 20136	118	177	170	11.2	0	0	Brown
19.	ICPL 20186	121	178	209	9.4	28	6	Brown
20.	ICPL 20205	128	189	220	10.2	0	0	Brown

Contd.

S.No.	Genotype	Days to flower	Days to mature	Plant height (cm)	100-seed weight (g)	Wilt %	Sterility mosaic%	Seed colour
21.	ICPL 20118	120	182	165	10.5	8	0	White
22.	ICPL 20126	119	180	172	12.2	0	0	Brown
23.	ICPL 20137	130	187	195	11.4	0	0	White
24.	ICP 7086	135	198	220	9.8	35	0	White
25.	ICPL 20117	130	191	198	10.3	0	0	Brown
26.	ICPL 20176	107	180	182	10.9	0	0	Brown
27.	ICPL 20177	121	181	190	8.4	9	0	White
28.	ICPL 20201	126	185	245	9.7	56	78	Brown
29.	ICP 10650	134	186	105	9.3	35	6	Brown
30.	ICP 8094	153	200	228	7.7	15	10	White
31.	ICPL 99044	131	185	185	10.5	0	0	White
32.	MA 3	137	190	198	9.2	22	0	Brown
33.	MA 6	154	204	220	9.3	13	0	Brown
34.	MA15	150	196	210	11.1	29	0	Brown
35.	ICP11376	136	204	170	8.5	0	0	Purple

Table 11: Stability of fertility restoration recorded in multilocation trials

S.No	Restorers ICP/ICPL No.	2010 Locations (4)	2009 Locations (3)	2008 Locations (3)	Mean
1.	ICPL 87119	96	85	97	93
2.	ICPL 20093	100	93	81	91
3.	ICPL 20096	89	97	95	94
4.	ICPL 20098	92	88	96	92
5.	ICPL 20104	99	99	93	97
6.	ICPL 20106	96	91	97	95
7.	ICPL 20107	95	88	91	91
8.	ICPL 20108	91	100	97	96
9.	ICPL 20111	95	93	95	94
10.	ICPL 20112	92	89	94	92
11.	ICPL 20116	98	91	87	92
12.	ICPL 20120	87	78	98	88
13.	ICPL 20123	100	96	95	97
14.	ICPL 20125	96	96	94	95
15.	ICPL 20127	95	96	88	93
16.	ICPL 20128	94	95	97	95
17.	ICPL 20129	95	95	81	90
18.	ICPL 20136	93	98	97	96
19.	ICPL 20186	76	96	91	88
20.	ICPL 20205	100	97	99	99
21.	ICPL 20118	91	80	-	85.5
22.	ICPL 20126	97	87	-	92
23.	ICPL 20137	100	100	-	100
24.	ICP 7086	100	98	-	99
25.	ICPL 20117	96	93	-	94.5
26.	ICPL 20176	93	98	-	95.5
27.	ICPL 20177	95	100	-	97.5
28.	ICPL 20201	98	100	-	99
29.	ICP 10650	92	98	-	95
30.	ICP 8094	95	100	-	97.5
31.	ICPL 99044	100	100	-	100
32.	MA 3	99	100	-	99.5
33.	MA 6	99	98	-	98.5
34.	MA15	100	91	-	95.5
35.	ICP 11376	95	91	-	93

Table 12: List of male-sterility maintainers and their important traits

S.No.	Genotype	Days to flower	Days to maturity	Plant height (cm)	100 seed weight (g)	Wilt %	Sterility mosaic %	Seed colour
Early Maturing								
1.	ICPL 11335	53	95	115	8	NA	NA	Brown
2.	ICP 14425	76	127	150	9.8	86	64	Brown
3.	ICP 14857	80	115	90	9.1	71	7	Brown
4.	ICP 16172	73	138	110	10.4	NA	NA	Brown
5.	ICP 14849	66	107	70	8.9	100	9	Brown
6.	ICP 10915	68	115	75	5.4	79	14	Brown
7.	ICP 10907	68	110	50	6.4	88	13	Brown
8.	ICPL 98011	66	112	145	8.7	81	38	Brown
	Mean	68.6	114.9	100.6	8.3			
	Sem(±)	2.9	4.6	12.7	0.6			
Medium Maturing								
9.	ICP 28	81	128	127	10.2	68	32	Brown
10.	ICPL 20282	98	148	185	10.3	42	50	Brown
11.	ICPL 20286	98	145	174	10.5	86	7	White
12.	ICPL 20288	102	158	185	11.2	67	17	White
13.	ICPL 20287	105	158	170	10.7	12	16	White
14.	ICPL 99050	123	175	225	11.1	0	0	Brown
15.	ICPL 20093	127	183	283	12	8	0	Brown
16.	ICPL 20099	127	184	292	14.7	5	0	Brown

Contd.

S.No.	Genotype	Days to flower	Days to maturity	Plant height (cm)	100 seed weight (g)	Wilt %	Sterility mosaic %	Seed colour
17.	ICPL 20094	129	185	280	10.6	0	0	Brown
18.	ICPL 20176	114	162	198	10	0	0	Brown
19.	ICPL 99052	123	178	235	11.9	0	0	Brown
20.	ICPL 118	103	146	132	13.7	0	2.2	Brown
21.	ICPL 96058	120	177	220	10.5	0	0	Brown
22.	ICP 5529	104	158	190	8.4	91	36	Brown
23.	ICPL 96053	128	184	198	10.5	0	4	White
	Mean	112.1	164.6	206.3	11.1			
	Sem(±)	3.7	4.6	13.0	0.4			
Late Maturing								
24.	ICP 14085	142	193	190	13.2	20	20	Brown
25.	ICPL 20092	148	198	140	9.6	23	0	White
	Mean	145.0	195.5	165.0	11.4			
	SE	3.0	2.5	25.0	1.8			
	Total Mean	100.9	151.2	169.2	10.2			
	Combined Sem(±)	5.4	6.2	13.2	0.4			

SE-Standard error of mean

4.5 Search for new cytoplasm sources

For a sustainable hybrid breeding program, it is essential that the breeding material has sufficient diversity with respect to both cytoplasm and nuclear genomes. The nuclear diversity is essential to breed high yielding hybrids and it has been discussed in Section 4.4. The importance of cytoplasmic diversity is also high in view of the fact that each hybrid plant will have the same cytoplasm and if any gene that is susceptible to any lethal disease is present in mitochondria, and if that disease comes, then all the plants will be affected. This was experienced earlier in the USA when gene for a devastating leaf blight disease was found in the cytoplasmic genome; and it caused almost total destruction of maize crop. In pigeonpea, a total of eight cytoplasm sources are available (Table 2) but all of them are not useful. Hence, efforts to search the new cytoplasm sources that can induce male sterility should continue.

4.6 Breeding disease resistant hybrids

In pigeonpea two diseases are of prime importance. These are *Fusarium* wilt (Fig 21) and sterility mosaic (Fig 22). A third disease, called *Phytophthora* blight (Fig 23 a & b) is also on increase but its incidence is limited to the poorly drained fields. At present, a number of resistance sources are available in pigeonpea germplasm for both wilt and sterility mosaic diseases; while for *Phytophthora* so far no reliable resistance source is available. Further, in most pigeonpea growing areas, both wilt and sterility mosaic diseases are prevalent that cause severe damage each year. Therefore, breeding involves for resistance to both the diseases simultaneously; and a number of genotypes with high levels of resistance are available.

For hybrid breeding also, the same approach of resistance breeding should be followed. However, for breeding hybrids, it is essential to understand the genetic nature of resistance so that appropriate breeding strategies can be developed. A review on this subject (Saxena and Sharma, 1990) revealed that the resistance to both the diseases was conferred by a pair of recessive alleles. Recently, Saxena *et al.* (2012) reported a dominant gene for resistance to wilt and this will be a very useful in breeding wilt resistant hybrids. This is because the hybrids involving as both resistant × resistant as well as resistant × susceptible parental lines will exhibit resistance to this disease.

The most important asset in breeding disease resistant hybrids is the availability of disease screening facility. This has been successfully accomplished at more than one location in India. In this dual disease screening nursery, wilt inoculums is added each year by incorporating wilted plants collected from different fields. For sterility mosaic disease, infector row technology is used. The details are

Fig. 21: ***Fusarium*** **wilt**

Fig. 22: Sterility mosaic disease (SMD)

Fig. 23a: ***Phytophthora*** **leaf blight**

Fig. 23b: ***Phytophthora*** **stem blight**

available in Nene (1988). Each year the incidence of diseases is monitored by growing susceptible check for each disease.

The following steps are suggested for breeding wilt and sterility mosaic disease resistant hybrid parents:

- Identify a reliable disease screening nursery with susceptible checks recording >95% disease incidence.
- Confirmation of disease resistance of the male sterile lines and testers by sowing the potential parents in the disease nursery for at least two seasons. To ensure high success, an attempt should be made to make crosses in the disease nursery using the resistant plants. Such crosses can be made after studying the diseases reaction. In this case, the pollinations should be done on the regenerated flush of flowers.
- Grow the experimental hybrids and take observations on fertility restoration on each plant. Identify maintainers and restorers among the testers.

- Repeat crosses which expressed >90% fertility restoration.
- Grow the experimental hybrids in the form of a trial in disease free fields and in addition single row of each hybrid in disease nursery for monitoring disease incidence.
- Multiply seed of the selected testers under self-pollination.
- Select hybrids showing disease resistance, high yield, and good seed traits; multiply their seed for multi-location testing.

5

Yield Assessments of New Hybrids

5.1 Seed back up programs

In hybrid breeding program, testing of new hybrid combinations is a continuous process and in most cases, the availability of crossed (hybrid) seed for their annual evaluation of selected hybrids is an important issue. Since, the selection of hybrids for advancement is based on yield data, one year will be required to produce seed for testing in the subsequent season. This means for each cycle of advancement, one valuable season will be consumed in their seed production. To avoid this situation and enhance the testing program at a fast pace, an aggressive hybrid seed production program should be followed. This can be achieved by visually selecting relatively more number of hybrids based on their flowering time, plant vigor, disease resistance, and early pod load about 4-5 weeks before harvesting. During this period, the new seed of the selected hybrids can be produced manually. This operation is facilitated by the perennial nature of pigeonpea plant, which can flower for extended periods of 40-50 days; that allows completion of desired number of pollinations of the target crosses in the same cropping season. One trained person can pollinate about 300 buds in 6-7 hours; and with 20% success, about 150 seeds can be produced. Experience has shown that, with good planning and crop management, this program can be achieved with ease.

5.2. Pollen sterility

The phenotypic expression of fertility restoration is not always uniform across the genotypes and environments. It may vary from zero (maintainer) to 100% (restorers). During the evaluation of new hybrids or backcrosses, such variations are not uncommon, particularly in early generations. Hence, it is essential to study pollen viability in each and every hybrid plant. This will ensure a rapid advance in the breeding populations. The pollen viability test is done by harvesting fully grown but still closed floral buds from different parts of the plants. The anther column is removed from the buds and squashed on a glass slide and drench it with 2% aceto-carmine solution. In each slide 3-5 microscopic fields should be examined using 10X magnification. In each slide counts should be taken for fertile (stained) and sterile (empty/ unstained) pollen grains. The male sterile hybrids are backcrossed with their respective testers to develop new male sterile (A-) lines. The fertile hybrids can be repeated for their agronomic evaluation in the subsequent season.

Fig 24 a: Plant with fully fertile pollen

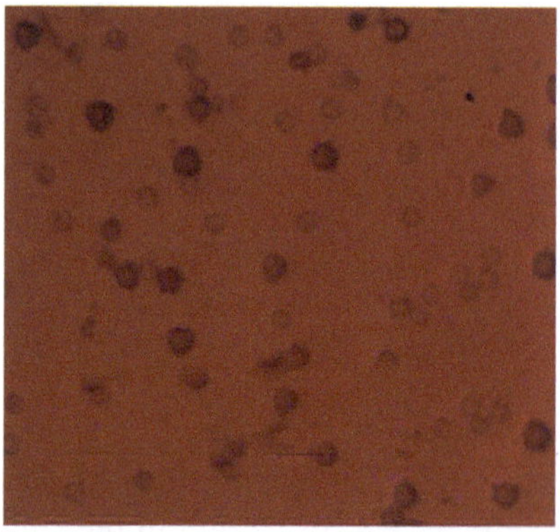

Fig 24 b: Plant with partial fertile pollen

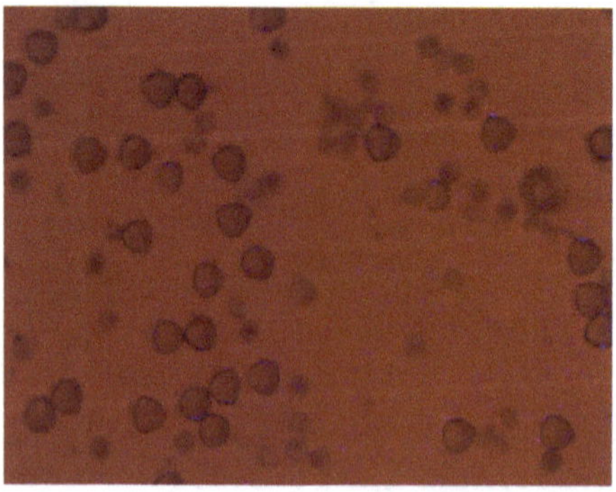

Fig 24 c: Plant with partial sterile pollen

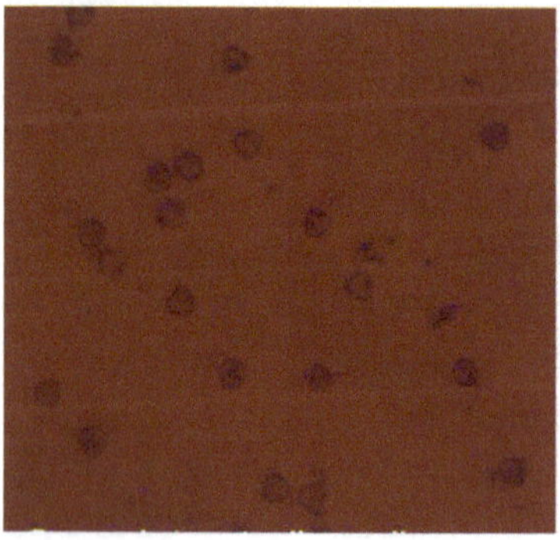

Fig 24 d: Plant with complete sterile pollen

5.3. Observation nursery

The evaluation of the first generation experimental hybrids is important to group the testers in to male sterility maintainers and fertility restorers. To achieve good results, it is important that the experimental hybrids with limited crossed seeds are evaluated properly with good care. Before sowing it would be good if the crossed seed lot is grouped on the basis of maturity of the parents; and in the field early, medium, and late groups should be planted with recommended density in different plots. This will facilitate good insect management and other field operations. The number of hybrid rows needed for planting will depend on the availability of the crossed seeds, which may differ from cross to cross.. Each hybrid and its tester should be sown side-by-side to allow their relative assessment for important parameters. In hybrid row each plant should be tagged for male sterility and fertility. The assessment for yield and key yield components should be done visually. The hybrids with predominantly male sterile plants indicate that their testers did not carry dominant fertility (Fr) gene; and hence be classified as 'maintainers'. In contrast, the testers of the fertile hybrids should be classified as 'fertility restorers'. The decision to retain the hybrids segregating for male sterility should be taken by the concerned breeder on the basis of the need, frequency of fertile/sterile plants, and expression of hybrid vigor.

5.4. Station trials

This is the real first field test of the new hybrids using new restorers. Each entry should be evaluated at least in two replications but three replications are always preferred. Always try to assess at least 15 plants for disease resistance. In this testing also strict observations should be made on fertility restoration and those confirming the last season's data should be observed carefully for yield and other traits; and hybrids with at least 30% superiority should be selected for second stage of station trials and if possible go for at least two locations. Only one or two selected hybrids with not less than 30% advantage and disease resistance should be promoted to multi-location testing under coordinated trials. The concerned breeder must plan well to multiply the seed of hybrids and their parents to continue three years mandatory assessment.

5.5 All India coordinated trials

The Indian Institute of Pulses Research, based at Kanpur provides an excellent opportunity to test the new genetic materials under diverse environments through its strong network spread over the entire country with standard plot size and replications. In order to identify pigeonpea cultivars for specific regions or for wider adaptation, the newly bred hybrids are tested with control cultivars. This is a multi-layered evaluation system and the new material needs to perform exceptionally good to qualify for the next level of testing. To make this testing a success and to generate meaningful data, it is essential that high quality of seed in sufficient quantities is produced by the breeders and made available for evaluation (Table 13).

Table 13: Grain yield (kg ha^{-1}) of Pigeonpea hybrid ICPH 2671 in Central Zone in All India Coordinated trials 2007

Genotype	SDAU, SK Nagar	MPKV, Rahuri	ZARS, Khargone	Mean	% Superiority over
Advance Hybrid Trial (AHT)					
ICPH 2671	2937	2590	3029	2852	-
Maruti (C)	2169	2660	2180	2336	22
Asha (C)	1833	2833	1764	2143	33
BSMR 736 (C)	2534	3611	2176	2773	3
Co 6 (C)	2944	3254	2125	2774	3
Initial Hybrid Trial (IHT)					
ICPH 2671	2934	2368	1527	2276	-
Maruti (C)	1493	2872	354	1573	45
Asha (C)	1931	2268	1550	1916	19
BSMR 736 (C)	1760	1604	1533	1632	39
Co 6 (C)	2546	2299	1081	1975	15
Mean of AHT & IHT					
ICPH 2671	2936	2479	2278	2564	-
Maruti (C)	1831	2766	1267	1955	31
Asha (C)	1882	2551	1855	1996	28
BSMR 736 (C)	2147	2608	1649	2135	20
Co 6 (C)	2745	2777	1603	2375	8

5.6 On-farm evaluations

This can be considered as an ultimate test of new hybrid and hence appropriate plans should be made in advance to produce the required seed quantity of high standards. Under this evaluation program a number of representative farmers' fields should be selected with a minimum size of one acre. This should be divided approximately into two halves; one for the test genotype and another for the control (the best local variety). Both the plots should receive the same recommended inputs and cultural treatments. To avoid complicated interactions, the evaluation of the two genotypes should be done under sole cropping. It is important to record observations on traits like maturity, uniformity, disease and insect reactions and seed yield. For comparison of relative productivity of the two genotypes, the recommended standard procedure of crop cutting should be adopted (Table 14).

5.7. High yield potential of hybrids

Under good management, pigeonpea hybrids have recorded yields as close as 4000 kg /ha (Table 14). This is a very positive indication about the yield potential of hybrids. The next generation hybrids bred with improved parents are likely to cross the bar of 5000 kg/ha.

Table 14: High yields (around 4000 kg l/ha) of Pigeonpea hybrid recorded by some farmers in Maharashtra

Location	Hybrid yield	Control yield	% Hybrid Advantage
Salod	3960	2040	94
Nimgaon	3950	2470	60
Kothoda	4670	3560	31
Tamoli	3890	2280	71

5.8. Grain quality and nutritional tests

Any new material produced by the breeders should meet the standards and parameters of traders and millers for efficient marketing. Besides this, it should also meet the expectations of both rural and urban consumers with respect to taste, cooking quality parameters, aroma and shelf life. Also, it should meet the minimum nutritional parameters such as protein percent and its digestibility and minerals etc (Table 15).

Table 15: Seed quality parameters of Pigeonpea hybrid ICPH 2671

Characters	ICPH 2671
Seed colour	Dark brown
100 seed mass (g)	11.4-11.8
Dal recovery %	76.49
Processing losses %	5.96
Water absorption(g/g)	2.14
Protein content (%)	19.86
Dal taste	Acceptable

6

Characterization of Elite Hybrids

Fig. 25: ICPH 2671 in field

Fig. 26: ICPA 2043 A (Female parent of ICPH 2671)

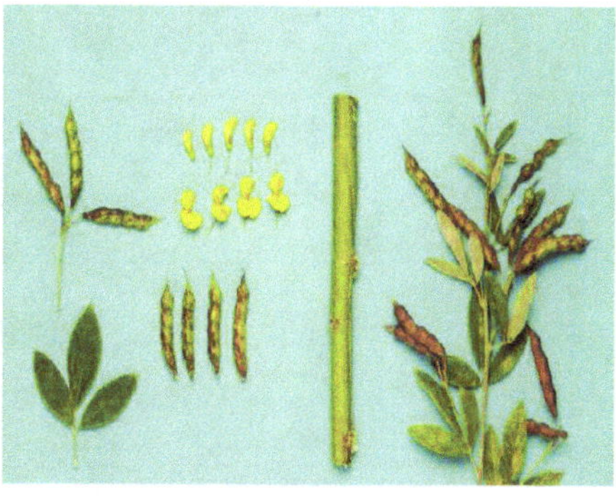

Fig. 27: ICP 2671 R (Male parent & restorer of ICPH 2671)

Table 16: Important traits of Pigeonpea hybrid ICPH 2671: ICPA 2043 (CMS line), and ICP 2671 R (Restorer line)

Characters	ICP 2043 A	ICP 2671 R	ICPH 2671
Days to flower	110-115	122-128	119-121
Days to maturity	158-165	180-185	168-176
Plant height (cm)	170-180	215-225	238-260
Growth habit	Indeterminate	Indeterminate	Indeterminate
Branching pattern	Spreading	Spreading	Spreading
Stem colour	Green	Green	Green
Leaf shape	Lanceolate	Lanceolate	Lanceolate
Flower colour	Light yellow	Yellow with light streaks	Yellow with red streaks
Pod form	Cylindrical	Cylindrical	Cylindrical
Pod colour	Green	Green with brown streaks	Purple
Seeds pod^{-1}	3.5-4.2	3.7-4.0	3.8-4.1
Seed colour	Brown	Brown	Dark brown
100 seed mass (g)	11.2-11.8	10.7-12.3	11.4-11.8
Disease resistance to		Wilt	Wilt

7

Comparison of Pigeonpea with Rice Hybrids

To overcome the challenge of popularizing pigeonpea hybrids in the country, it is considered fair enough to compare the hybrid rice with respect to their seed production technology, yield advantages and profitability to the farmers and seed producers. Since both pigeonpea and rice are grouped under self-pollinated crops, this comparison, in the views of authors, may help in encouraging hybrid pigeonpea seed producers and those involved in marketing of hybrid seed.

Rice is a highly self-pollinated crop and earlier hybrids were not possible because it was essential that large quantities of hybrid seeds are produced easily and economically. In China, the birth place of hybrid rice, this impossible looking task was accomplished. To launch this program, the Chinese scientists decided to alter the pollination system of rice. They forced the exertion of panicle and extended the duration of floret opening in the female plants by spraying a plant hormone GA_3. For mass cross pollination, they spread pollen grains of male plants over the treated female rows by shaking the flowering male plants with the help of a rope or stick. This operation suspended the pollen grains in the air and settled on the exerted female flowers for inducing artificial cross-pollination. In the beginning, this approach looked impossible to implement considering high seed rate (30 kg/ha) and poor seed set by cross-pollination. But in spite of heavy odds, the Chinese persisted and made it to happen by slowly improving the technology. They developed and implemented the national master plan at research, production, and distribution levels; and this made the things to happen with positive results. Some important comparative features of both the systems are given in Table 17.

Table 17: Comparative statement of seed and commercial crop production of rice and pigeonpea

Parameter	Rice*	Pigeonpea**
Hybrid Seed Production		
Seed rate	20 - 25 kg/ha	5-6 kg/ha
Seed production cost	40 - 50 Rs/kg	7-8 Rs /kg
Benefit : cost ratio	-	6.1
Seed-to-seed ratio	1: 50	1: 200
Commercial Hybrid Crop		
Base cultivar yield (kg/ha)	5000-6000	1000-2000
Hybrid yield (kg/ha)	5750-7000	1800-3500
Hybrid advantage (kg/ha @15-20%)	750-1000	800-1500
Net extra returns from hybrid cultivation (Rs/ha)	5000 – 6000	40,000- 75,000

* Rice data source: Virmani (1996); Ahamad and Siddiq (2013)
**Pigeonpea data source: different farmers of Maharashtra and Andhra Pradesh

8

Strategies for Future: Meeting the Challenge of Imports Through Hybrids

With the advent of hybrid technology, it is natural to assess its potential benefits at the national level. To meet the domestic requirement of pigeonpea, about 50.000–60,000 metric tons of pigeonpea is imported annually from Myanmar and Africa. In this communication, an attempt has been made to look into the possibility if we can meet the entire domestic needs of pigeonpea through the adoption of hybrid technology. It is a big question but the goal is not impossible. The estimates shown in Table 18 show that if only 2% of total pigeonpea area can be brought under hybrids, the target can be met with ease. This will require seed production in only 700-800 hectares.

Table 18: Estimated calculation for promotion of hybrids to stop imports of pigeonpea

S.No.	Content	Estimates
1.	Estimated annual imports	60,000 tons
2.	Total area under pigeonpea	40 lakh ha
3.	Area that can be brought under hybrids (2%)	80,000 ha
4.	Hybrid seed required for 80,000 ha @ 8 kg/ha	640 tons
5.	Area required to harvest 640 tons of seed	650 ha
6.	Add 10-15% failures of seed plots	750 ha
7.	Breeder Seed needed for 1000 ha of sowing	8000 kg
8.	Area for Breeder Seed production	10-15 ha
9.	Extra production through hybrids 80,000 x @ 0.8 t/ha	64,000 tons

9

Moving Forward

The hybrid breeding program in pigeonpea is unique and the first among the food legume. The concept of hybrid in this crop was conceived in 1974; and it is based on a small window of natural out-crossing present in the crop. This natural out-crossing in pigeonpea is mediated by a range of flying insects, which forage on large bright colored flowers of the crop in search of nectar and in the process affect the cross-fertilization. In the last 40 years a lot of basic and applied research has been successfully completed and over 50 research papers have been published in national and international journals of repute. A coordinated effort of scientists of different institutions has led to the release of two hybrids, one for central and another for peninsular zone of India.

The advantage of hybrids in terms of enhanced yields has been demonstrated in farmers' fields in different states. Pigeonpea hybrid technology is now established and it is easier and cheaper than rice. Seed production is no more an issue, but selection of production site is critical. Seed-to-seed ratio in hybrid pigeonpea is higher (1: 200) than rice (1:50). A good hybrid seed chain can now be developed. A breakthrough in pigeonpea productivity is possible.

These hybrids have demonstrated that in pigeonpea exploitable hybrid vigor is available in sufficient quantities and the investment for developing high yielding hybrids is highly justified for benefiting farmers, who have waited for a very long time for a breakthrough in the productivity of this crop. The road ahead is not easy and scientists, research managers and policy makers have to work together with seed sector to make this technology a reality. In this context the following research and development areas will require attention:

- Nuclear and cytoplasmic diversification of hybrid parents.
- Breeding environment and cropping system specific hybrids.
- Classify hybrid parents in heterotic groups for planning specific hybrids and improvement of parents.
- Identify hybrid parents with high GCA and SCA.
- Regular screening of hybrids and parents for disease resistance.
- Appropriate cataloguing and storage of parents and hybrids.
- Standardize seed quality testing procedures of hybrids at different levels.
- Develop region specific agronomic package for maximizing hybrid yields.
- Develop hybrids with yielding ability of 6-7 tonnes/ha under high input system.
- Identify ideal seed production locations.
- Prepare map of seed production sites for different districts to guide seed producers.
- Update field plot techniques for optimizing hybrid seed yields in different areas.
- Characterize elite hybrid parents using morphological and genomic tools.
- Develop seed production standards in collaboration with seed research division.
- Incorporate morphological markers in parental lines to maintain genetic purity.
- Develop links with public and private seed sector.
- Conduct and publish basic and applied research to improve the hybrid technology.
- Conduct training programs regularly for efficient technology transfer.
- Popularize cultivation of hybrids among farmers.

References

Ariyanayagam, R.P., A.N. Rao and P.P. Zaveri, (1995). Cytoplasmic male-sterility in interspecific matings of cajanus. *Crop Sci.* **35:** 981-985.

Bhatia, G.K.; Gupta, S.C.;Green, G.M. and D.Sharma (1981). Estimates of natural cross-pollination in *Cajanus cajan*(L.) Millsp.Several Experimental Approaches. *Proc. Intl. Workshop on Pigeonpeas* **2**: 129-136. ICRISAT 15 19 December, 1980.

Bohra A, Saxena RK, Ganesh BN, Saxena KB, Byregowda M, Rathore A, Kavikishor PB, Cook DR and Varshney RK. (2012). An intra-specific consensus genetic map of pigeonpea [*Cajanus cajan* (L.) Mill sp.] derived from six mapping populations. *Theoretical and Applied Genetics* **25** (6):1325-38.

Bohra, A., A. Dubey, R.K.Saxena,R.V.Penmetsa,K.N Poornima,N,Kumar,A.D. Farmer, G. Srivani, H.D. Upadhyaya, S.R. Ramesh, D.Singh, K.B. Saxena, P.B. Kavi Kishore,N.K. Singh,C.D.Town,G.D. May, D.R. Cook, and R.K. Varshney (2011). Analysis of BAC end sequences(BESs) and development of BES-SSR markers for genetic mapping and hybrid purity assessment in pigeonpea (*Cajanus* spp.). *BMC Plant Biol.***11**:56-70.

Byth, DE, Saxena, KB and Wallis, ES. (1982). A mechanism for inhibiting cross fertilization in pigeonpea [*Cajanus cajan* (L.) Millsp.]. *Euphytica* **31**: 405-408.

Dalvi, V.A.; Saxena, K.B. and Madrap, I.A. (2008). Fertility restoration in cytoplasmic - nuclear male-sterile lines derived from three wild relatives of pigeonpea. *J. Hered.* **99**: 671-673.

Dalvi, VA and Saxena, KB. (2009). Stigma receptivity in pigeonpea (*Cajanus cajan* (L.) Millsp.) *Ind J. Genet.* **69**: 247- 249.

De, D.N. (1974). Pigeonpea. *In:* J. B. Hutchinson (Ed.) Evolutionary studies on world crops. Cambridge University Press. Cambridge. pp. 79-87.

FAO. (2012) www.faostat.org

Hanson MR and Bentolila S (2004) Interactions of mitochondrial and nuclear genes that affect male gametophyte development. *The Plant Cell* **16**:154-69.

Howard, A.;G/L/C/Howard and R.A.Khan (1919). Studies in the pollination of Indian crops.I.Memories,Deptt. Agric. India (Botanical Series) **10:**195-200.

Jelena, S., Snezaana, M D., Zorica, P. and Milomir F. (2007). Characterisation of maize inbred lines based on molecular markers, heterosis and pedigree data. *Genetika*. **39**: 335-363.

Jones, HA and Emsweller, SL. (1937). A male sterile onion. *Proc. Ann. Soc. Hort. Sci.* **343**: 582-585.

Kadam,B.S.; Kulkarni,R.M. and R.M.Patel(1945).Natural crossing in *Cajanus cajan* (L.) Mill sp. In the Bombay Deccan. *Indian J.Genet.* **5**:60-62.

Kaul MLH (1988). Male sterility in higher plants. *In:* Frankel R, Grossman M, Maliga P, Riley R, eds. Monographs of Theoretical & Applied Genetics 10. Heidelberg: Springer-Verlag. 1005 pp.

Khan, T.N. (1973). A new approach to the breeding pigeonpea (*Cajanus cajan* Millsp.) Formation of composites. *Euphytica* **22**:373-377.

Kumar, R.V. and Saxena, K.B. (2001). First report of wind pollination in pigeonpea. *Ind. J. Genet.* **61** (1): 279-280.

Levings, C. (1993). Thoughts on cytoplasmic male sterility in CMS-T maize. *Plant Cell,* **5**:1285-1290.

Lu, XG., Zhang, G. Maruyama, K. and Virmani, S.S. (1994). Current status of two line method of hybrid rice breeding. *In:* Hybrid rice technology; New developments and future prospects (S.S. Virmani ed.). pp 37-50. International Rice Research Institute. Manila. Philippines.

Mahta, D.N. and B.B.Dave (1991). Studies in *Cajanus indicus* Memories, Deptt. Agric. India (Botanical Series) **19:**1-25.

Malikarjuna, N. and Saxena, K.B. (2005). A new cytoplasmic nuclear male-sterility system derived from cultivated pigeonpea cytoplasm. *Euphytica*, **142**(1-2):143-148.

Mallikarjuna, N.D. Jadhav, and P. Reddy (2006). Introgression of *Cajanus platycarpus* genome into cultivated pigeonpea genome. *Euphytica*. **149**:161-167.

Mallikarjuna N, Saxena KB, Jhansi Lakshmi, Varshney RK, Sandhya Srikanth and Deepak Jadhav (2012). Differences between *Cajanus cajan* (L.) Millspaugh and *C. cajanifolius* (Haines) van der Maesen, the progenitor species of pigeonpea. *Genetic Resources and Crop Evolution* **59**: 411-417.

Naresh, V., Yamini, KN., Rajendrakumar, P., and Dinesh, KV. (2009). EST-SSR marker- based assay for the genetic purity assessment of safflower hybrids. *Euphytica* **170**: 347- 353.

Nene YL (1988). Multiple disease resistance in grain legumes. *Annual Review of Phytopathology*. **26:** 203-217.

Pathak GN. (1970). Red Gram. *In:* Pulse Crops of India. Indian Council of Agricultural Research. New Delhi, India. pp 14–53.

Rachit K. Saxena, Eric von Wettberg, Hari D. Upadhyaya, Vanessa Sanchez, Serah Songok, Kulbhushan Saxena, Paul Kimurto and Rajeev K. Varshney (2014). Genetic Diversity and Demographic History of *Cajanus* spp. Illustrated from Genome-Wide SNPs. *PLOS One* DOI: 10.1371/journal.pone.008856.

Reddy MV, Sharma SB and Nene YL (1990). Pigeonpea: disease management. *In:* Nene YL, Hall SD, Shiela VK (eds.) The pigeonpea. CAB International, pp. 303–348.

Richey FD. (1922). The experimental basis for the present status for corn breeding. J. Amer. Agron. **14:** 1-17.

Sawargaonkar SL, Madrap IA and Saxena KB (2012). Study of inheritance of fertility restoration in pigeonpea lines derived from *Cajanus cajanifolius*. *Plant Breed*. **131**: 312-314.

Saxena, K.B.(2008).Genetic improvement of pigeonpea-a review. *Tropical Plant Biology* **1**:159–178.

Saxena, K.B.(2014). Temperature-sensitive male-sterility system in pigeonpea. *Current Science*, **107**(02): 277-281.

Saxena, K.B., Durga, B.K. and L.Singh (1987). Ineffectiveness of wrapped flower in inhibiting cross-fertilization in pigeonpea. *Euphytica* **36**:295-297.

Saxena,K.B. and Kumar,R.V.(2013) Pigeonpea. *In*: Hybrid seed production in field crops. Kalyani Publishers, New Delhi-12. India. pp 213-240.

Saxena KB, Kumar RV, Srivastava N and Shiying B (2005). A cytoplasmic-genic male-sterility system derived from a cross between *Cajanus cajanifolius* and *Cajanus cajan*. *Euphytica*, **145**: 291-296.

Saxena, KB., Kumar, RV., Saxena, RK., Sharma, M., Srivastava, RK., Sultana, R., Varshney, RK., Vales, MI. and Pandey, S. (2012). Identification of dominant and recessive genes for resistance to *fusarium* wilt and their implication in breeding hybrids. *Euphytica* **188**: 221-227.

Saxena K.B. Kumar, R.V. Srivastava N, Shiying B. (2005). A cytoplasmic-genic male sterility system derived from a cross between *Cajanus cajanifolius* and *Cajanus cajan*. *Euphytica*, **145**:291-296.

Saxena K.B. (2013). A novel source of CMS in pigeonpea derived from *Cajanus reticulatus*. *Indian J. Genet.* **73:**259-263.

Sexena K.B. and R.V. Kumar (2003). Development of a cytoplasmic nuclear male sterility system in pigeonpea using *C. scarabaeoides* (L.) Thouars. *Ind. J. Genet.* **63:**225-229.

Saxena K.B., RV. Kumar, N. Srivastava, and S. Bao, (2005). A Cytoplasmic-nuclear male-steriltiy system derived from a cross between *Cajanus cajanifolius* and *Cajanus cajan*. *Euphytica.* **145:**289-294.

Saxena, K.B., Laxman Singh and M.D. Gupta (1990). Variation for natural out-crossing in pigeonpea. *Euphytica* **46:**143-148.

Saxena, K.B. Sultana, R. Mallikarjun, N., saxena, R.K., Kumar. R.V., Swargaonkar, S.L. and Varshney R.K. (2010). Male sterility systems in pigeonpea and their role in enhancing yield. *Plant Breeding*, **129:**125-134.

Saxena, K.B. (2014). Temperature-sensitive male-sterility system in pigeonpea. *Current science*, **107:**277-281.

Saxena, KB., Kumar, RV., Tikle. AN., Saxena MK., Gautam, VS., Rao, SK, Khare, D., Chauhan, YS., Saxena, RK., Varshney, RK., Reddy, BVS., Sharma, D., Reddy, LJ., Green, JM., Faris, DG., Mula, M., Sultana, R., Srivastava, RK., Gowda, CLL. and Sawargaonkar, SL. (2013). ICPH 2671- The world's first commercial food legume hybrid. *Plant Breeding*. **132**: 479-485.

Saxena, KB., Kumar, RV and Bharathi, M. (2014). Studies on fertility restoration in pigeonpea. *Euphytica.* **196**: 127-135.

Saxena, KB.and Sawargaonkar,SL.(2014). First information on heterotic groups in pigeonpea (*Cajanus cajanus* (L.) Millsp.) *Euphytica*. **200** (2):187-196.

Saxena, KB., Sharma, D. and Green, JM. (1976). Pigeonpea rationing – An aid to breeders. *Tropical Grain Legume Bulletin* **4**: 21.

Saxena, KB. and Sharma, D. (1990). Pigeonpea Genetics. *In:* The Pigeonpea. (Nene,YL,Hall, Sd, and Shiela, eds). CAB Intl. Wallingford. UK. 137-158.

Saxena, K.B., Laxman Singh and M.D. Gupta (1990). Variation for natural out-crossing in pigeonpea. *Euphytica* **46** (2): 143-148.

Saxena, KB., Sultana, R., Mallikarjuna, N., Saxena, RK., Kumar, RV., Sawargaonkar, SL. and Varshney, RK. (2010). Male sterility systems in pigeonpea and their role in enhancing yield. *Plant Breeding* **129**: 125-134.

Saxena KB, Sultana R, Saxena RK, Kumar RV, Sandhu JS, Rathore A, and Varshney RK (2011a). Genetics of fertility restoration in A_4 based diverse maturing hybrids in pigeonpea [*Cajanus cajan* (L.) Millsp.]. *Crop Science* **51**: 574-578.

Saxena KB, Vales MI, Kumar RV, Sultana R and Srivastava RK (2011b) Ensuring genetic purity of pigeonpea hybrids by incorporating a naked-eye polymorphic marker in A and B lines. *Crop Science* **51**: 1564-1570.

Saxena, K B and Kumar, R V and Saxena, R K and Sharma, M and Srivastava, R K and Sultana, R and Varshney, R K and Vales, M I and Pande, S (2012). Identification of dominant and recessive genes for resistance to *Fusarium* wilt in pigeonpea and their implication in breeding hybrids. *Euphytica*, **188** (2). pp. 221-227.

Saxena, MK., Saxena, U., Saxena, KB., Khandelkar, VS., and Sultana, R. (2011).Profitability and production cost of hybrid pigeonpea seed. *Electronic Journal of Plant Breeding* **2**: 409-412.

Saxena RK., Prathima C., Saxena KB., Hoisington DA., Singh NK. and Varshney RK. (2010). Novel SSR markers for polymorphism detection in pigeonpea (*Cajanus* spp.). *Plant Breeding* **129**:142-148 .

Shi, M. (1981). Preliminary report of later japonica natural -2 lines and application. *Hubei Agriculture Science*. **7**: 9-12.

Shull, GF. (1908).The composition of a field of maize. Report of American Breeders' Association. **4:** 296-301.

Singh AK. and Gopalkrishnan S. (2013). Male sterility restoration. *In:* Hybrid seed production in field crops. Kalyani Publishers, New Delhi-12. India. pp 17-46.

Stephens JC and Holland, RF. (1954). Cytoplasmic male sterility for hybrid sorghum seed production. *Agronomy Journal*. **46**: 20-23.

Tikka SBS, Panwar LD and Chauhan RM (1997). First report of cytoplasmic genic male sterility in pigeonpea (*Cajanus cajan* (L) Millsp.) through wide hybridization. *GAU Res. J.*, **22:**160-162.

Van der Maesen LJG (1990). Pigeonpea: origin, history, evolution and taxonomy. *In:* Nene YL, Hall SD and Sheila VK (eds). The Pigeonpea. Wallingford: CAB International, pp 44–87.

Veeraswamy, R., G.S. Palaniswamy and R. Rathinaswamy (1973). Natural cross pollination in *Cajanus cajan* (L) Millsp. *Lablab niger* Medikus. *Madras Agric. J.* **60:** 9-12.

Wang Z, Y Zou, X Li, Q Zhang, L Chen, H Wu, D Su, Y Chen, J Guo, D Luo, Y Long, Y Zhong, and Liu Y G (2006). Cytoplasmic male-sterility of rice with boro ii cytoplasm is caused by a cytotoxic peptide and is restored by two related ppr motif genes via distinct models of m-rna silencing. *Plant Cell* **18:** 676-387.

Wanjari, K.B., A.N. Patil, P. Manapure, J.G. Manjaya and P. Manish, (1999). Cytoplasmic male-sterility in pigeonpea with cytoplasm from *Cajanus volubilis*. *Ann.Plant Physiol.* **13:**170-174.

Wilsie,C.P.and M.Takahashi (1934). Natural crossing in pigeonpea. *J.Agri.Res.* **49**(10):923-925.

Yuan, LP. (1987). Strategy conception of hybrid rice breeding. *Hybrid Rice*. **1:** 1-4.

MIX

Papier | Fördert
gute Waldnutzung

FSC® C083411

Zeitfracht Medien GmbH
Ferdinand-Jühlke-Straße 7
99095 Erfurt, Deutschland
produktsicherheit@kolibri360.de